1 MONTH OF
FREE
READING

at

www.ForgottenBooks.com

By purchasing this book you are eligible for one month membership to ForgottenBooks.com, giving you unlimited access to our entire collection of over 1,000,000 titles via our web site and mobile apps.

To claim your free month visit:

www.forgottenbooks.com/free65968

ISBN 978-1-5282-6642-0
PIBN 10065968

SHORT COURSE

IN

QUANTITATIVE

CHEMICAL ANALYSIS.

BY

JOHN HOWARD APPLETON, A.M.,

Professor of Chemistry in Brown University.

———

SEVENTH EDITION.

———

SILVER, BURDETT & CO., PUBLISHERS.

NEW YORK.   BOSTON.   CHICAGO.

1895.

PREFACE

TO THE REVISED EDITION.

In this little book, the Author endeavors to provide a laboratory guide suitable for the use of beginners in Quantitative Analysis.

1. The course prescribed is so short as to be capable of study from beginning to end within the time allotted to an ordinary course of practical Chemistry.

2. On the other hand, it has been the aim to have the work include such a number of exercises as will enable the student not only to become acquainted with the methods of determining all the most frequently-occurring elements, but also to study a variety of type-processes.

3. A considerable number of explanatory notes have been provided. It is believed that these will be helpful to both teacher and pupil; to the latter these notes will not only be instructive—they will help him to work understandingly rather than mechanically.

4. The plan of the book contemplates regular recitations by the student. The author deems this to be a matter of great importance. He believes that sooner

or later every one admits the advantage of being able to express his thoughts in clear and appropriate language; and again, one can hardly find a better means of testing the accuracy of his knowledge of a given subject than that afforded by the opportunity of stating his views before a capable judge of the matter.   Hence experienced educators agree in attaching a high importance to the habit of recitation as a means of increasing facility and accuracy both in thought and in expression.

In the present edition an alphabetical table of atomic weights has been added (p. 182).   It contains the results of recent revisions of these important numbers.   It also gives the numbers for newly-discovered elements, such as gallium, germanium, samarium, and scandium.

Throughout the book a large number of minor changes have been made, in accordance with recent advances in the science of chemistry.

BROWN UNIVERSITY, 1892.

CONTENTS.

1 *

6 CONTENTS.

EXERCISES.

APPENDIX.

HINTS TO TEACHERS.

1. Commence work by showing the pupil how to weigh with the balance that he is expected to use. Now let him weigh, in the presence of the teacher, a small portion of Alum, the substance first tested. Next let him go to his desk and there perform, as described in the book, the experiments necessary to the analysis of the substance in question.

2. Require of each pupil a stated amount of work. Of course this amount must vary with the time the student spends in the laboratory. Thus, some students cannot be expected to make more than three tests per week, while others will easily have time for six.

The time required for the faithful study of the work as laid out in this book, is estimated at about two hundred hours of laboratory work and about twenty hours of oral recitation.

3. Require the student to enter in the Table given on pages 10, 11, and 12 the results of each of the three tests that he makes by each of the processes described in the book.

4. At such time as suits the arrangements of the teacher, there should be a recitation upon the text of the book.

5. It is desirable—though not strictly necessary—that each student should have, as his own property, a set of accurate weights, a platinum filter-cone, a platinum crucible. When a student graduates, he can readily sell such apparatus to a member of the succeeding class.

6. The arrangement of the matter in the book is such that the teacher can easily omit portions of it, if he thinks it best to do so.

STUDENT'S RECORD OF ANALYSES.

REPORT OF QUANTITATIVE WORK,

By ....

from .... *to* ....

	Theoretical per cent.	Per cent. by Tests.	
ALUMINIUM, in Alum, as Al_2O_3,	5.970	....	1 2 3
ANTIMONY, in Tartar emetic, as Sb_2S_3,	36.177	....	4 5 6
ANTIMONY, in Tartar emetic, as Sb_2O_4,	36.177	....	7 8 9
ANTIMONY, in Tartar emetic, by $BaSO_4$,	36.177	....	10 11 12
ARSENIC, in Arsenious oxide, as $Mg_2As_2O_7$,	75.771	....	13 14 15
ARSENIC, in Arsenious oxide, as As_2S_3,	75.771	....	16 17 18
BARIUM, in Barium chloride, as $BaSO_4$,	56.204	....	19 20 21
BISMUTH, in Bismuthyl nitrate, as	72.705	....	22 23 24
BROMINE, in Potassium bromide, as $AgBr$,	67.172	....	25 26 27

10

	Theoretical per cent.	Per cent. by Tests.
CALCIUM, in Calcium carbonate, as Ca O,	40.040	.... 28 .... 29 30
CARBON DI-OXIDE, in Calcium carbonate, loss by ignition,	43.944	.... 31 .... 32 33
CARBON DI-OXIDE, in Calcium carbonate, by Johnson's apparatus,	43.944	.... 34 .... 35 36
CARBON DI-OXIDE, in Calcium carbonate, by Scheibler's apparatus,	43.944	.... 37 .... 38 39
CHLORINE, in Sodium chloride, gravimetric,	60.616	.... 40 .... 41 42
CHLORINE, in Sodium chloride, volumetric,	60.616	.... 43 .... 44 45
CHROMIUM, in Potassium di-chromate, as $Cr_2 O_3$,	35.410	.... 46 .... 47 48
COPPER, in Cupric sulphate, by iron,	25.392	.... 49 .... 50 51
COPPER, in Cupric sulphate, by electrolysis,	25.392	.... 52 .... 53 54
COPPER, in Cupric sulphate, as Cu O,	25.392	.... 55 .... 56 57
IRON, in Ammonio-ferrous sulphate, as $Fe_2 O_3$,	14.278	.... 58 .... 59 60
IRON, in Ammonio-ferric sulphate, by Potassium permanganate,	11.617	.... 61 .... 62 63
IRON, in Ammonio-ferrous sulphate, by Stannous chloride,	14.278	.... 64 .... 65 66
LEAD, in Lead nitrate, as $Pb SO_4$,	62.519	.... 67 .... 68 69
LEAD, in Galena, as $Pb SO_4$,	86.583	.... 70 .... 71 72

	Theoretical per cent.	Per cent. by Tests.
MAGNESIUM, in Magnesium sulphate, as $\overline{Mg}_2\,P_2\,O_7$,	9.772	.... 73 .... 74 75
MERCURY, in Mercuric chloride, as HgS,	73.854	.... 76 .... 77 78
NICKEL, in Ammonio-nickelous sulphate, by electrolysis,	14.714	.... 79 .... 80 81
NICKEL, in Ammonio-nickelous sulphate, as $Ni\,O$,	14.714	.... 82 .... 83 84
NITROGEN, in Ammonium chloride, by Distillation,	26.217	.... 85 .... 86 87
NITROGEN, in Ammonium chloride, as $(N\,H_4)_2\,Pt\,Cl_6$,	26.217	.... 88 .... 89 90
NITROGEN, in Potassium nitrate, Pelouze's process modified,	13.875	.... 91 .... 92 93
PHOSPHORUS, in Hydro di-sodium phosphate, as $\overline{Mg}_2\,P_2\,O_7$,	8.671	.... 94 .... 95 96
POTASSIUM, in Potassium chloride, as $K_2\,Pt\,Cl_6$,	52.419	.... 97 .... 98 99
SILICON, in Felspar, as SiO_2,	30.393	....................100 ....101 102
SILVER, in United States coin, as $AgCl$,	90.000	....................103 ....104 105
SULPHUR, in Cupric sulphate, as $Ba\,SO_4$,	12.857	....................106 ....107 108
TIN, in Stannous chloride, volumetrically by $K_2\,Cr_2\,O_7$,	52.474	....................109 ....110 111
ZINC, in Zinc sulphate, as ZnO,	22.653	....................112 ....113 114

QUANTITATIVE ANALYSIS.

CHAPTER I.

GENERAL REMARKS.

Gravimetric Analysis and Volumetric Analysis.

A QUANTITATIVE test is usually performed either by the gravimetric method or by the volumetric method; in some cases both methods are combined in one and the same test.

GRAVIMETRIC ANALYSIS.

A gravimetric test usually involves the following processes:

First, weighing the substance to be tested;

Second, dissolving this portion of substance and preparing it for the subsequent operations;

Third, precipitating the substance in the form decided upon;

Fourth, filtering and washing the precipitate in order to separate it from the other matters present;

Fifth, drying the precipitate, usually upon the filter-paper on which it was collected;

Sixth, burning the paper and the precipitate separately, the paper first;

Seventh, weighing the substance left after burning;

Eighth, making the necessary calculations.

Each of these steps must be carefully taken, in ordet that the final result may be a satisfactory one.

The Weighing.

This subject is discussed in the next chapter.

The Dissolving.

Some of the principles connected with dissolving are so well known and so general in their application that they need not be discussed in this book; others are so special that they are best referred to in connection with the processes to which they apply. In any case, the analyst who gives careful consideration to the amount and proper method of application of water or of other appropriate solvent he is to use, will be amply repaid in the greater speed and greater accuracy thereby attained.

The Precipitation.

A given chemical element is usually capable of forming more than one insoluble compound; hence the analyst has usually some range of choice as to the special form in which he will precipitate and weigh the element he is studying. The compound selected should fulfil as many as possible of the following conditions:

(*a*) It should be insoluble in the menstruum present; if it is granular and compact, rather than flocculent, it is the more easily and safely washed in the later steps of the process.

(*b*) It should be non-volatile; this prevents loss during the burning of the filter-paper.

(*c*) It should not undergo any change of weight upon exposure to the air; this reduces the difficulties of weighing.

(*d*) It should be of known molecular constitution;

this enables the analyst to calculate the amount it contains of the particular element sought.

(*e*) It is well if it has a high molecular weight; then the element to be estimated forms a smaller proportion of such quantity of the compound precipitated as is necessarily lost in the operations of the analysis.

The Filtering and Washing.

The Paper.—The filter-paper used should be sufficiently stout and porous; it should also be as free as possible from mineral matter. An excess of mineral matter not only tends to the contamination of certain filtrates, it also renders the analyst more liable to error in making his subtraction for the average filter-ash.

The manufacturer of filter-paper selects a vegetable fibre that naturally contains but little mineral matter; he also usually subjects it to treatment with acids and alkalies in order to remove as much as possible of that mineral matter necessarily present.

Sometimes the analyst purifies his own filter-papers before use by soaking them in dilute acid, and subsequently washing and drying them.

The Funnel.—The sides of the funnel should form an angle of exactly 60 degrees. The reason is obvious. If we consider the way in which a plain filter is folded, we perceive that only half of it is utilized; this half forms an inverted cone, the circumference of whose base is one-half the circumference of the original filter when flat. Hence the slant-height of the cone — the radius of the original filter when flat — is equal to the diameter of the base of the same cone. Thus a plane section through the axis of the cone affords an equilateral triangle, and hence its angles measure each 60 degrees.

The neck of the funnel should be long and rather narrow, and it is best to have the lower extremity cut off perpendicular to the axis. In this condition there is a tendency to the accumulation of the filtrate in the neck; when at length this filtrate runs out, its fall assists filtration by producing a slight suction upon the liquid yet in the filter.

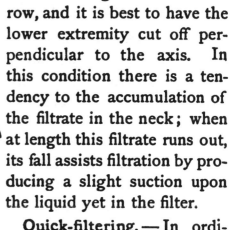
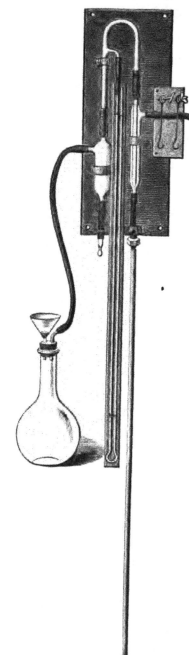

Quick-filtering. — In ordinary filtering the liquid passes through by reason of its own weight. But the last portions of such liquid form such a short column that it exerts but little downward pressure; further, the resistance of the paper and of thick and abundant precipitates makes the operation tedious. Of course the atmospheric pressure is exerted upon the upper surface of the liquid; but this affords no help, since it is balanced and neutralized by like upward pressure exerted in the neck of the funnel.

The natural and simple idea of applying suction — that is, of removing the atmospheric pressure from below — was

FIG. 1.—Quick-filtering Apparatus.

not made practicable until Bunsen suggested the use of

two devices; these are the Sprengel water-pump, for producing the necessary exhaustion, and the platinum cone, for sustaining the point of the filter.

The apparatus for quick-filtering consists in general of two portions, the filtering apparatus and the suction apparatus. Both parts of the appliance are capable of much modification, and a variety of forms have been devised. The following description points out the general principles involved.

The filtering apparatus consists of the funnel, which must be correctly shaped; the platinum cone, which may be perforated to advantage; the filter-paper, which must be folded "plain," must fit closely to the funnel and over the platinum cone; the filter-flask, which must be supplied with a cork in one of whose perforations the funnel-neck fits air-tight, the other perforation being connected with the suction apparatus.

The suction apparatus consists essentially of a chamber of glass or of metal supplied with three openings. Into the first opening enters the supply of water; from the second opening the waste-pipe conveys the water away; at the third (the side opening) is attached the air-tube that conducts the rarefied air from the filter-flask to the waste-pipe that carries away the mingled air and water.

It is very evident that the force that does the work of exhaustion is derived from the falling water; and further, that the force of the falling water is derived either from a powerful "head" of a high tank (*e. g.*, the reservoir of a city), or else—as in the original Bunsen arrangement— from the downward fall of a considerable column of water in the long waste-pipe.

In this latter form of pump the following conditions must be complied with:

2*   B

(*a*) The waste-pipe must be larger than the supply-pipe.

(*b*) The waste-pipe must be of such small diameter that the water supplied to it shall not trickle down on one side, but shall adhere to it on all sides, and thus — falling like a series of disks — shall sweep the air down before it.

(*c*) The waste-pipe must be of considerable length, so that the column made up of the sum of the water-bubbles may be able to fall with sufficient rapidity. But the space within the waste-pipe being insufficiently supplied with water (see *a*), the air in the system — including that in the filter-flask — continually expands and flows toward the waste-pipe; there, it is swept down in bubbles by continual flow of the deficient supply of water.

A case might occur in which there would be little downward flow in the waste-pipe, except when the column of water should aggregate 34 feet in height; in such a case there might be produced a sort of water barometer with an imitation, at its summit, of the Torricellian vacuum, and now the atmospheric pressure at the lower opening of the waste-pipe would steadily support the column of water in this pipe and maintain it at a height of 34 feet. It hardly needs to be said that, in practice, besides some leakage of air through the spaces about the filter-paper, there are usually other conditions sufficient to determine available suction with a waste pipe of much less than 34 feet vertical fall.

Advantages of the Filter-Pump.

By use of the filter-pump the filtering and washing of a precipitate are accomplished,

First. More rapidly.

Secondly. By smaller relays. and so more thoroughly (see Johnson's *Fres. Quant. Anal.*, p. 67).

Thirdly. The filtrate is kept small. The attainment of the latter result is a double advantage: it lessens the loss due to the solubility of most precipitates; it saves time and material in those numerous cases in which a filtrate must be evaporated before receiving subsequent treatment.

The Drying.

In most cases the filter with its precipitate should be thoroughly dried before its ignition. Some precipitates inclose water in their pores and retain it with great tenacity; if they are incompletely dried, the ignition changes the residual water into steam, with sudden explosion and loss of particles of the precipitate.

The Ignition.

The process of ignition contemplates not only the complete expulsion of everything in the filter-paper and precipitate that is combustible, but further, it seeks the attainment of this result without injury to the precipitate itself.

Much advantage accrues from separating the precipitate from the paper and then burning the paper before the precipitate is ignited. Of course, if the separation could be made perfect, the paper might be at once cast aside; but the latter is usually burned, in order to preserve such small portions of precipitate as still adhere to it. The previous separation renders combustion less tedious, since many precipitates manifest a tendency to protect the paper from the oxygen of the air; it also spares the mass of precipitate from subjection to decomposing influences of the burning paper. Although the small portion of precipitate that adheres to the paper frequently suffers decomposition, yet it is often possible to restore it to its original condition. Thus in case of minute portions of Silver chloride, the chlorine lost is easily re-supplied; the same is true of the oxygen that is lost when traces of barium sulphate are reduced to barium sulphide by the burning of the paper to which it had adhered.   (See p. 65.)

For the purposes of ignition, platinum crucibles have many advantages over porcelain; they accomplish a great saving of time by burning organic matters with great rapidity (a result that appears to be due in part to the occlusion of oxygen by the platinum); the metal dishes also cool very quickly in the desiccators. With moderate care, platinum dishes last so long (and are ultimately salable as scrap platinum) that they are really cheaper than porcelain.

At the beginning of a combustion filter-paper burns at a gentle heat better than at a very strong one; the latter seems to liberate from the cellulose or woody fibre $(C_6H_{10}O_5)$ of the paper, a difficultly combustible carbon resembling graphite.   Toward the end of the combustion an increase of heat is attended with good results.

Calculations.

Three calculations are usually necessary.

RULE FIRST.—To make out the *theoretical per cent.* of the element sought in the problematical substance to be analyzed, work out the following proportion:

The molecular weight of the original substance, under examination,

is to

the atomic or the molecular weight of the element sought,

as

one hundred per cent.

is to

the theoretical per cent. of the element sought.

RULE SECOND.—To find the *amount by weight* of any element or substance that is contained in any given weight of a compound, work out the following proportion:

The molecular weight of the compound,
is to
the atomic or the molecular weight of the element or substance sought,
as
the gross weight of the compound in question,
is to
the gross weight of the element or substance sought.

RULE THIRD.—To make out the *per cent. by test*, work out the following proportion:

The gross weight of the original substance taken for analysis,
is to
the gross weight of the element sought (see Rule Second),
as
one hundred per cent.
is to
the per cent. by test.

VOLUMETRIC ANALYSIS.

Volùmetric methods are generally quick methods, and are of great value where many analyses of the same kind have to be performed.

A volumetric test usually involves the following processes:

First. Preparing one or more standard solutions.

Second. Weighing the substance to be tested.

Third. Dissolving the weighed portion of substance, and preparing it for the subsequent operations.

Fourth. Titration.

Fifth. Making the necessary calculations.

The Standard Solutions.

A standard solution is one containing the reagent in a strength that is known with a high degree of exactness. There are at least two ways in which this strength may be fixed.

One method depends upon dissolving an exactly-weighed amount of reagent in a carefully-measured amount of liquid. Thus, if the analyst has exactly five grammes of Silver nitrate in five hundred cubic centimetres of solution, he can easily determine the exact amount of common salt that one cubic centimetre of the silver solution will precipitate.

Another method depends upon making the standard solution of a strength approximating to that desired, and then afterwards, by a properly devised test, discovering the exact strength. Thus, if the analyst has a solution of Silver nitrate of unknown strength, he can learn its true relations to common salt by careful trial upon an exactly-weighed amount of the pure salt. This latter method of standardizing the solution is often to be preferred.

Standard solutions are capable of variation in strength. (*a*) Rise of temperature may expand the solution so as to decrease its apparent strength; (*b*) fall of temperature may contract the solution so as to increase its apparent strength; (*c*) evaporation of liquid from the solution may produce a real increase of strength; (*d*) accidental addition of liquid to the solution may produce a real decrease of strength.

Weighing.

This subject is discussed in Chapter II.

Dissolving.

After dissolving the substance to be tested, it is often best to dilute the solution to a certain measured bulk, five hundred cubic centimetres or one litre; several separate fifths or several separate tenths may then be tested by the standard solution, and so give the results the confirmatory value of several different tests.

Titration.

This is the process of carefully dropping the standard solution into the solution of the substance to be tested. The standard solution is drawn from a graduated tube called a burette, and the quantity added must be just sufficient to have its full chemical action upon the substance tested.

The completion of the reaction is generally marked by a striking change of color in some special substance or reagent suitable for the purpose and called an indicator; thus, litmus is the indicator in acidimetry and akalimetry, and a mixture of Potassium iodide and starch is the indicator in a number of volumetric processes involving oxidizing operations.

The analyst must give careful thought and attention to such conditions as may give rise to errors of measurement in titration. Some conditions tend to make the readings too high, some tend to make them too low.

FIG. 2.—Mohr's Burette-clamp.

Too high reading may arise from not filling the delivery tube *below the clamp* with standard solution at the beginning of the test; and the same error may result from a small bubble of air inadvertently left in the same place, but finally forced out by the current of the solution and thus counted for its volume of standard solution.

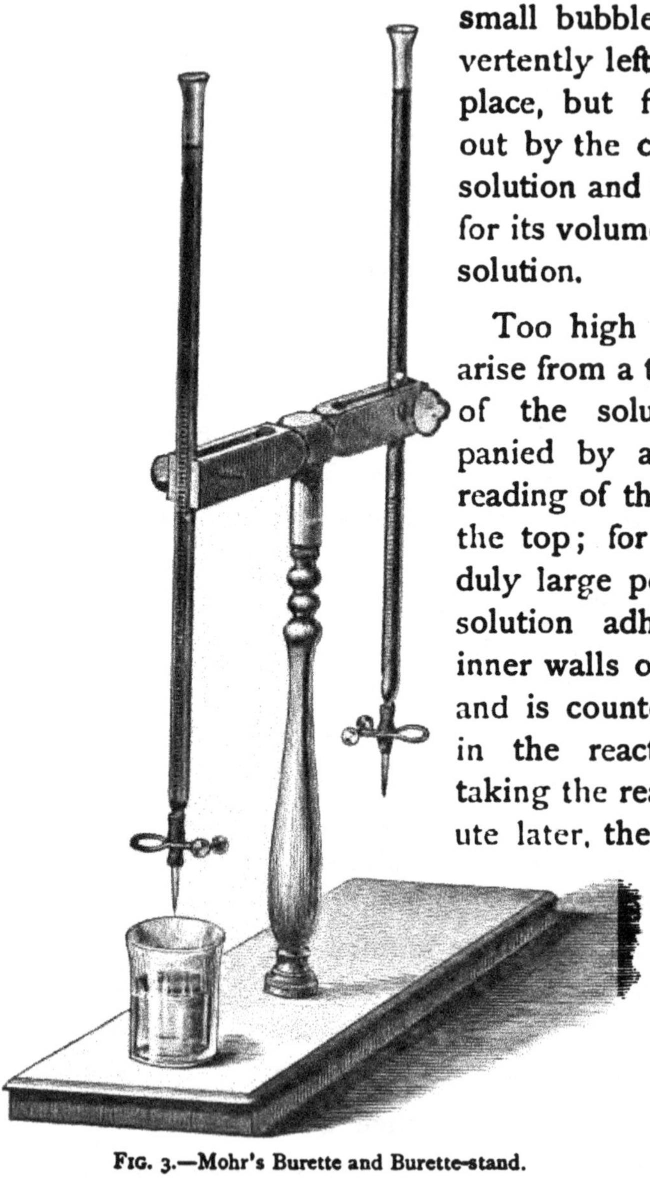

Too high reading may arise from a too rapid flow of the solution accompanied by an immediate reading of the position of the top; for thus an unduly large portion of the solution adheres to the inner walls of the burette and is counted as if used in the reaction. Upon taking the reading a minute later, the error is apparent, for the drip has now raised the top of the column.

Too high reading may arise in some cases from too slow a flow of the standard solution. This is particularly the case with solutions that are liable to change upon exposure to the air. Thus some portion of

FIG. 3.—Mohr's Burette and Burette-stand.

the standard solution (changed in fact by the influence of the air) is counted as used in combining with the substance under examination.

Too high reading may come from introducing the standard solution into a fresh burette that has just been emptied of water; in this case the water adhering to the walls of the burette, slightly dilutes the first filling of standard solution. This last-mentioned source of error is avoided by rinsing the burette before use with a small amount of the standard solution. Of course these rinsings are thrown away.

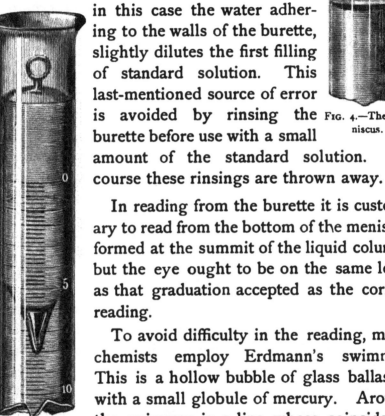

FIG. 4.—The Meniscus.

FIG. 5.—Erdmann's Swimmer.

In reading from the burette it is customary to read from the bottom of the meniscus formed at the summit of the liquid column; but the eye ought to be on the same level as that graduation accepted as the correct reading.

To avoid difficulty in the reading, many chemists employ Erdmann's swimmer. This is a hollow bubble of glass ballasted with a small globule of mercury. Around the swimmer is a line, whose coincidence with the graduations of the burette aids the eye of the analyst.

The Calculations.

These are best discussed under the special cases.

3

CHAPTER II.

THE CHEMICAL BALANCE.

The Chemical Balance, its construction, its theory, conditions of its sensibility and its accuracy. Methods of weighing. Rules for weighing.

Construction of the Balance.

THE Chemical Balance consists essentially of the following parts:

(1) The beam with its knife-edges,

(2) the agate-plates,

(3) the stirrups and the pans,

(4) the arresting apparatus,

(5) the glass case.

(1) The beam has three knife-edges—the central one, called the fulcrum, has its sharp edge directed downward, the other two have their sharp edges directed upward. All the knife-edges are firmly attached to the beam. When the balance is not in use the beam is supported by two strong arms of metal. When in action the beam is capable of free motion, for then its central edge rests upon the central agate-plate, and the end edges support the agates of the stirrups.

(2) The central agate is firmly attached to the column. The end agates rest freely upon the end edges.

26

(3) The stirrups are the frames from which the pans are suspended.

(4) The arresting apparatus takes various forms. In its simplest form it consists of two arms, which may be made to lift the whole beam, so as to raise the centre

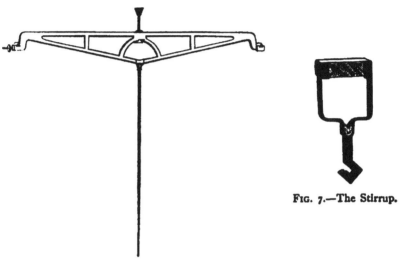

FIG. 7.—The Stirrup.

FIG. 6.—The Balance-Beam, showing Knife-Edges and Index.

knife-edge from its agate-plate; by this means the centre knife-edge and its agate-plate are spared, when not in actual use.

In a better form of arresting apparatus, the mechanism lifts not only the beam, but, in addition, the stirrups; thus the end knife-edges and plates are spared.

Sometimes pan-stoppers are provided; these are light arms that rise under the pans and offer a slight but equal resistance to any swaying of the latter.

(5) The glass case shields the balance, when in use, from currents of air; it also protects the balance from moisture and from corrosive gases.

The case may be supplied with a small beaker containing either concentrated Sulphuric acid or solid Calcium chloride to absorb moisture, or containing Quick-lime or

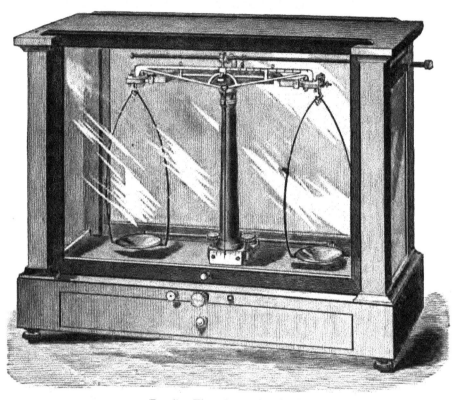

Fig. 8.—The Chemical Balance.

solid potassium hydroxide to absorb at once moisture and acid vapors.

Theory of the Balance.

The sensibility of the balance and its accuracy, are qualities that are closely connected but are readily distinguished. To secure these qualities in a high degree, the balance must be constructed in accordance with certain clearly defined principles.

1. The centre of gravity of the system must be below the fulcrum, but

2. It must be as near the fulcrum as possible.

3. The three knife-edges must be in the same straight line.

4. The beam must be rigid.

5. The beam must be as light as possible.

6. It is well to have the two end knife-edges equidistant from the fulcrum; that is, to have the arms of the beam of equal length.

7. The knife-edges must be sharp.

First. If the centre of gravity of the system is above the fulcrum, the balance is in the condition of unstable equilibrium — hence the beam is easily overset.

If the centre of gravity is at the fulcrum, the balance is in neutral equilibrium — hence even when the weights in the opposite pans are equal, the beam may rest in a position of deviation, so that the operator is deceived as to the true relation of the weights.

If the centre of gravity is below the fulcrum, the condition is that of stable equilibrium — the whole system is that of a short but wide pendulum. In this case, when the weights in the opposite pans are equal, the long index will come to rest at the Zero of the ivory scale. But when the weights are unequal, this fact will be indicated by the proper deviation. Under these conditions, the heavier pan descends, and the point of the index describes a certain arc whose angle is called the angle of deviation. This angle is a measure of the sensibility of the balance; the smaller the excess of weight required to produce deviation to a given angle, the greater the sensibility of the balance.

3*

Second. Let us consider two cases—one in which the

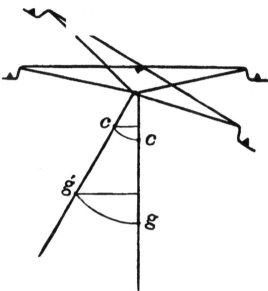

centre of gravity is far below the fulcrum, as at *g*, the other in which it is but a very small distance below, as at *c*. It is manifest that in the first case—in order to produce a given deviation—the centre of gravity must be moved through a greater vertical distance than in the second case; in other words, more

FIG. 9.—Diagram to illustrate different positions of the Centre of Gravity of the system supported by the Fulcrum.

work must be done. Of course, then, in the first case supposed, the balance will be less sensitive.

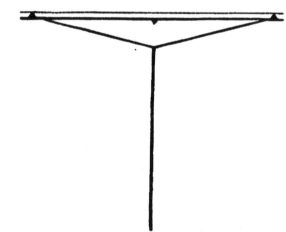

FIG. 10.—Diagram to illustrate a bad construction, in which the Terminal Edges are on a line above the central one.

Third. This requirement may be discussed under three cases:

(*a*) If the terminal edges are in a line above the central one, then the addition of a load to the pans may carry the centre of gravity up to the fulcrum, or even above it, and so give rise to the objectionable conditions referred to in the preceding paragraphs.

(*b*) If the terminal edges are in a line below the cen-

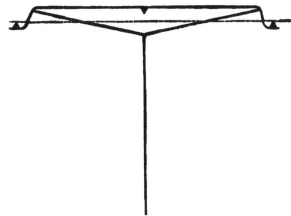

Fig. 11.—Diagram to illustrate a bad construction, in which the Terminal Edges are on a line below the central one.

tral one, then addition of loads to the pans will lower the centre of gravity— a condition already shown to be injurious to sensibility.

(*c*) If the central and terminal knife-edges are all in the same straight line, then the addition of loads to the pans tends to raise the centre of gravity— though the latter can never thus reach the fulcrum.

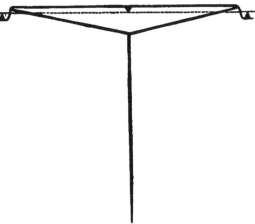

Fig. 12.—Diagram to illustrate a correct construction, in which the Central and Terminal Edges are all in the same straight line.

So far, here is a tendency

to increase the sensibility; but the thereby increased weight of the whole system makes this a theoretical rather than a real advantage, for, in fact, all balances are less sensitive with greater loads.

Fourth. Should the beam be flexible, the addition of weights to the pans will tend to lower the centre of gravity and so to lessen the sensibility. (See Second condition.)

Fifth. The less the weight of the system, the less the weight necessary to move it—that is, to turn the balance.

Sixth. For the most accurate weighing, the arms of the beam should be of equal length. When the arms are of unequal length, the analyst should uniformly place the substance weighed on the same side—say, on the left-hand pan; then though the weighings may all give

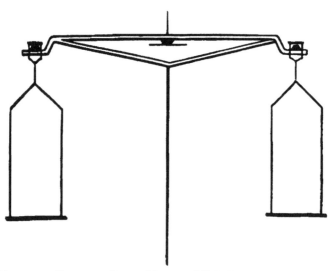

FIG. 15. Diagram to illustrate a Beam with very dull knife-edges, the Beam being in a horizontal position.

erroneous results, yet the error will represent a constant fraction, and the several weighings will still bear their true relations to each other.

Seventh. Suppose the knife-edges to be very sharp; then the lengths of the arms will maintain a constant value, even when the beam is inclined. Now suppose the edges to be very dull. Suppose the arms of the

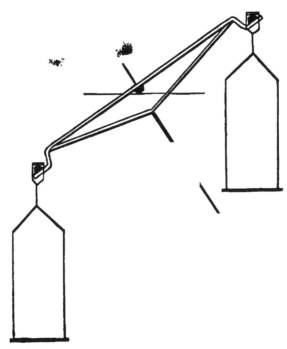

Fig. 14.—Diagram to illustrate a Beam with very dull knife-edges, the beam being in an inclined position. (The upper branch gains in length; the lower branch loses in length.)

beam to be of equal length. Represent this length by *x*. Represent the width of the edges as the same quantity called *y*. Now when the beam is inclined, as in the cut, the value of the right-hand arm is

$$x + \tfrac{1}{2}y + \tfrac{1}{2}y = x + y$$

while the value of the left-hand arm is

$$x - \tfrac{1}{2}y - \tfrac{1}{2}y = x - y$$

hence under these conditions the arms of the beam— when inclined—may differ by 2 *y*.

Position of the Balance.

The balance should be placed where it is not exposed to direct heat—either of the sun or of any artificial source of heat. With this in view, it is well to place the balance where the light is from the North.

The balance should be placed where it will not be subject to vibrations of the floor. With this in mind, the balance should be placed, if possible, on a shelf attached to a wall or pier based on the foundations of the building.

Weighing in the Air.

Whenever a body is immersed in a liquid or in a gas, the body experiences an upward pressure; this upward pressure is equivalent in amount to the weight of a quantity of the liquid or gas, equal in volume to the volume of the body immersed. Hence, when a body is weighed in the ordinary manner in a balance, both the body and the weights experience this upward pressure from the air. Now when the volume of the body is different from that of the weights—as is generally the case—they experience a different amount of upward pressure. Hence an error is here introduced. In most cases this error is not very great; but in the most accurate determinations it is avoided by reducing the weighings to vacuum—or better, by actually performing the operations *in vacuo.**

* W. Crookes : The Chemical News, XXIX., p. 14, The Atomic Weight of Thallium. H. W. Chisholm : Weighing and Measuring, p. 169.

Methods of Weighing.

First. **The Ordinary Method—Direct Weighing.**

This method is the one generally employed. It is capable of accuracy sufficient for all ordinary purposes; it is also expeditious. By it the substance to be weighed is generally (best always) placed in the left-hand pan, and the weights are then placed in the right-hand pan.

Second. **Gauss's Method.—Weighing by Reversal.**

The substance is weighed first in one pan, then in the other. If the two weighings give the same result, then —other conditions being correct—this result shows the true weight. If the results are different, the geometric mean of them is the true weight. (The geometric mean is the square root of the product of the two observed weights: $\sqrt{W \times W'}$.)

$x =$ True weight of object
$y =$ Ratio of shorter arm to longer arm
Apparent smaller weight, $W = \dfrac{x}{y}$
Apparent larger weight, $W' = xy$
$\dfrac{x}{y} \times xy = x^2$
whence $x = \sqrt{W \times W'}$

But it is evident that in most cases, the arithmetical mean — half the sum of the two observed weights — gives results that are sufficiently accurate.

Third. **Borda's Method.—Weighing by Substitution.**

In this method, the body to be weighed is first counter-balanced by placing small shot, or any other suitable and convenient material, in the opposite pan. The body itself is now withdrawn, and weights are added in its place until they produce equilibrium again. Of course the weights so added represent, with a high degree of accuracy, the weight of the body in question.

Rules for Weighing.

First. Take such a position that the Zero of the ivory scale is directly in front.

Second. Before weighing, carefully release the beam, and make sure, by trial, that the pans are in equilibrium when not loaded. (The beginner should never attempt to *adjust* a balance; in case of need, he should apply to the Professor in charge of the laboratory.)

Third. Never add anything to the pans nor beam, nor withdraw anything from them, except when the beam is supported upon its stops.

Fourth. Never weigh any substance while it is hot. Hot substances generate currents in the air, and so give rise to deceptive vibrations of the beam. Hot vessels must be cooled before weighing. But in the ordinary air, they may absorb moisture during cooling. To avoid errors from this source, hot vessels are cooled in desiccators. These are vessels in which a dry atmosphere is maintained by constant presence of concentrated Sulphuric acid or of Calcium chloride, or of some other drying agent.

Fifth. Substances to be weighed should not be placed directly upon the pans, but should be weighed in suitable vessels.

Sixth. Platinum vessels — especially large ones — should not be weighed immediately after wiping them; they should first be allowed time to attract their normal film of atmospheric air. (J. Lawrence Smith, Am. Chemist, V., p. 212.)

Seventh. The weights themselves must not be touched, except with the tweezers specially provided for handling them.

Eighth. In placing the weights on the pans, always arrange them in three definite lines — one line for the weights representing units, another line for tenths, a third line for hundredths. Thousandths are usually weighed with the *rider.*

Ninth. After the proper weights have been placed upon the pan, the greatest care must be taken to read and record them correctly. With this in view, make a record of the weights while reading from the vacant spaces of the weight-box; next, read the weights again, one by one, as they are transferred back to their places. The operator should learn — so as to know it perfectly —

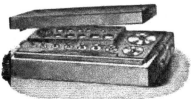

FIG. 15.—A Box of Weights.

the exact order of the series of weights he is using. This order is usually the following, or some slight variation of it, — 5, 2, 1, 1. The same order is usually preserved for large as well as for small denominations.

Tenth. Use public property as carefully as if it were your own.

4

CHAPTER III.

WEIGHTS AND MEASURES.

The Origin of Weights and Measures. The English System. The French or Metric System.

SPECIALLY prominent among the measures in most common use are those of extension and those of the force of gravitation. The measures of extension comprise linear, superficial, and cubical measures — the latter including various measures of capacity. The measure of the force of gravitation is called weight.

Primitive man naturally adopted standards of measurement of extension from parts of his own body — witness such measures as the hand-breadth, the cubit, the foot, and the fathom. (The pace, as a unit of itinerary measure, may be properly included in this group.)

Measures of capacity — it is easy to believe — originated with the use of a convenient shell or hollowed gourd, yet there is evidence that man early defined his measures of capacity in terms of his linear measures.

The origin of the measures of weight is obscure; they were probably derived from the weight of certain natural objects, as, for example, from dried grains from the middle of the ear.

But civilized man is not content until he has a system of measures, founded upon some invariable standard capable of accurate definition; and upon attaining a high degree of enlightenment, he seeks to increase to the utmost the accuracy and convenience of his metrological systems. In order to accomplish these ends, he must proceed in accordance with the business habits of his times as well as with its scientific knowledge.

Evidently two courses are open. One course is to correct and adjust an old system, harmonizing its various parts and placing the whole upon an accurate basis. This has been in general the policy of Great Britain.

Another course is to devise an entirely new system, complete and connected in all its parts. This method has been followed in France — the metric system being substituted in its entirety for the old measures of that country. (The empire of Germany has adopted the metric system.)

The English System.

For many centuries the English units of measure have been certain pieces of wood or metal, legally declared to be the standards, and carefully preserved as such.

But a law, which took effect in 1826, and continued nominally in force until 1855, made an entirely new departure. The law embodied an attempt to base the standards of measure upon the dimensions of some unchangeable objects or constants found in nature, to which reference could always be made in case of need. This law provided that in case of loss of the legal standard or injury to it, the yard must be obtained from the length of a pendulum vibrating seconds of mean time, in a vacuum, at the level of the sea, and latitude of London, the temperature being 62° Fahrenheit. The law stated the length of such a pendulum to be 39.1393 inches of the standard yard. In like manner it provided that the pound, if lost or injured, should be restored by a reference to the weight of a certain measured volume of distilled water; the act prescribed that the weight of one cubic inch of water is 252.458 grains of the weight then standard.

The general intent of this law was merely to make a new and more accurate and scientific definition of the then legal standards without altering their value. But it was soon discovered that the assigned length of the seconds pendulum and the assigned weight of the cubic inch of water were not correct. The errors arose from mechanical and physical difficulties in the way of absolutely exact experimental determinations of the natural constants referred to — difficulties that have not been wholly overcome even at the present day.

In 1834 occurred an accident for which it was originally supposed that ample provision had been made, but which perhaps was not contemplated as probable : the standard yard and the standard pound were ruined by a fire in the Parliament buildings. This accident gave rise to the appointment of a commission of eminent scientific men, who carefully examined the whole subject of weights and measures. This commission decided that it was advisable to reconstruct the lost standards by recourse to certain well-authenticated copies still existing, rather than by conforming to the law, which positively required the use of the seconds pendulum. The commission did, in fact, restore the lost standards from copies, and later the law was altered so as to conform to this method of restoration.

The present legal standards of Great Britain, therefore, are certain pieces of metal and certain vessels, of which accurate copies may be made.

The yard is the distance between two certain lines on a certain piece of bronze.

The pound is the weight *in vacuo* of a certain piece of platinum designated as P S (Parliamentary standard).

The gallon is the bulk of ten pounds, or 70,000 grains of water; it is believed to equal 277.274 cubic inches.

The bushel is the bulk of eight such gallons; it is believed to equal 2218.1907 cubic inches.

The Measures of the United States.

The English colonists in America depended at first upon weights and measures brought from England. But discrepancies gradually arising in the different States, attention of Congress was called to the whole matter of weights and measures, and at length a Senate resolution of March 3, 1817, directed Mr. John Quincy Adams (then Secretary of State) to report upon the present usage of this and of foreign countries and upon the best future course of the United States.

In 1821, four years later, Mr. Adams made an elaborate and learned report upon the points in question. While awarding high praise to the metric system as one which, "whether destined to succeed or doomed to fail, will shed unfading glory upon the age in which it was conceived and upon the nation by which its execution was attempted and has been in part achieved," yet he considered that it was best for the United States to adhere in the main to the English system.

That recommendation has been followed. But while the Constitution of the United States gives to Congress the power "to fix the standard of weights and measures," that body appears not to have passed any act directly establishing or adopting any particular system. It is true that the Secretary of the Treasury had certain standards constructed in 1832, assuming that the authority to do

4*

this properly belonged to his department, and later (in 1836) Congress appeared to sanction his acts by directing him to furnish standards of weight and measure to each State in the Union. In some cases, at least, the State laws have since provided that special standards then furnished or hereafter to be furnished by the United States shall be the legal standards of the States in question.

The original source of the standard yard of the United States appears to have been a certain brass scale 82 inches in length, deposited in the office of weights and measures in Washington. The temperature at which this scale was standard was 62° Fahrenheit, and the yard measure was between the 27th and the 63d inch marks of the scale. But the present standard yard appears to be a new bar obtained in London in 1856.

The standard avoirdupois pound is the weight of a certain piece of metal preserved as the standard; it is derived from a certain Troy pound of brass obtained in England in the year 1827.

The gallon is the bulk of 231 cubic inches of the standard yard. This is the so-called wine gallon of the earlier English system; it is one of six different gallon measures of unequal dimensions that have at different times been used in England. It is supposed to contain 58,372.1754 grains of water at 39.83° Fahr.

The bushel is the bulk of 2150.42 cubic inches of the standard yard. It is supposed to be equal to the bulk of 543,391.89 grains of distilled water.

The French or Metric System.

The metric system is so called because it has its basis in a measure of length called a *metre*. The system is of French origin, and is one of the products of the great revolution. In 1790, instead of attempting to harmonize the discordant systems then in use in France, it was proposed that an entirely new one be formed. A committee of scientific men was appointed for the purpose, and in 1793 the system recommended was provisionally adopted, and later (in 1798) the commission presented its report, together with a Platinum metre and a Platinum kilogramme.

The inventors of the system intended that the metre should be the forty-millionth of the terrestrial meridian. With this in view, a distance of about ten degrees on the meridian from Dunkirk, France, to Barcelona, in Spain, was carefully measured; thence the length of the quadrant was learned in terms of the measures then in use. Finally, the quadrant being divided into ten million parts, one of those parts was called a metre. But it is believed that the standard metre now in use is not the exact ten-millionth of even that quadrant selected. In common with other human undertakings the measurement of the meridian was not perfect. Later examination appears to have shown that the metre in use is about *one ten-thousandth* too short.

Whether this is so or not is a matter of little real consequence; it is not probable that in any event a re-measurement of the quadrant will be made for purposes of weights and measures. It is generally admitted that whatever the ideal metre may be, the metre *in fact* is the length of a certain piece of metal carefully preserved, and from which other metres must be derived by copying.

This measure of length (the metre) is multiplied and divided decimally, and from it superficial and cubical measure are directly derived.

The measure of weight is derived from the linear measure. Suppose a cubical box measuring one centimetre—one-hundredth of a metre—in each direction, its capacity being one cubic centimetre. Such a box filled with distilled water, at its temperature of greatest density, 4° Centigrade, would hold a certain quantity of water, whose weight would be the unit of weight to be called one gramme. (The exact method just suggested was not employed owing to its mechanical unfitness for precise measurements; the more practicable method used for fixing the weight of a given volume of water was by observing the loss of weight experienced by a certain bulk of metal when weighed immersed in water.)

The gramme is divided and multiplied decimally.

The names of metric measures are constructed systematically; thus for names of multiples of a given unit, Greek prefixes are attached to the name of the unit; for names of sub-divisions, Latin prefixes are employed.

The metric tables are given below.

Table of Metric Measures.

MEASURES OF LENGTH.

1 Millimetre	=	.001 of a metre	
1 Centimetre	=	.010 of a metre	
1 Decimetre	=	.100 of a metre	= about 4 inches.
1 Metre	=	1.000 metre	= 39.37 inches.
1 Decametre	=	10.000 metres	
1 Hectometre	=	100.000 metres	
1 Kilometre	=	1,000.000 metres	= about ⅝ of a mile.
1 Myriametre	=	10,000.000 metres	

SPECIAL MEASURES OF SURFACE (as of land).

1 Centiare	=	1 Square metre	= about 1⅕ square yards.
1 Deciare	=	10 Square metres	
1 Are	=	100 Square metres	
1 Hectare	=	10,000 Square metres	= about 2½ acres.

SPECIAL MEASURES OF VOLUME (of cord wood).

1 Decistere	=	.1 of a cubic metre
1 Stere	=	1.0 cubic metre
1 Decastere	=	10.0 cubic metre

MEASURES OF CAPACITY.

1 Millilitre	=	.001 of a litre
1 Centilitre	=	.010 of a litre
1 Decilitre	=	.100 of a litre
1 Litre	=	1.000 litre
1 Decalitre	=	10.000 litres
1 Hectolitre	=	100.000 litres
1 Kilolitre	=	1,000.000 litres

NOTE. 1 Litre $= \left\{ \begin{array}{l} \text{1 cubic decimetre} \\ \text{or 1000 cubic centimetres} \end{array} \right\} =$ about 1 quart.

MEASURES OF WEIGHT.

1 Millogramme	=	.001 of a gramme	= about $\frac{1}{65}$ of a grain.
1 Centigramme	=	.010 of a gramme	
1 Decigramme	=	.100 of a gramme	
1 Gramme	=	1.000 gramme	= about $15\frac{1}{4}$ grains.
1 Decagramme	=	10.000 grammes	
1 Hectogramme	=	100.000 grammes	
1 Kilo (gramme)	=	1,000.000 grammes	= about $2\frac{1}{5}$ lbs.
1 Tonneau	= 1,000.	kilos	= about 1 ton.

Advantages of the Metric System.

The metric system is used in quantitative analysis by chemists of all nationalities. It is also largely used by other scientific men. This use, not being compulsory, must be referred to certain real advantages possessed by the system. Of these, the following may be enumerated:

First. Its use of decimal relationships.

Second. The suggestiveness of the names it uses.

Third. The simplicity with which it expresses the relation between the weight of water and its volume, this

simplicity being readily communicated to any other substance whose specific gravity is known.

The wide, and yet increasing, use of the system in the
scientific literature of all nations is a fact that cannot be
overlooked.

Remark on the Metallic Standard.

From the account already given it is evident that the
metrical systems of the great commercial nations of the
world are based on metallic standards.   But a metal
standard is itself changeable in length—not only by
influence of heat and cold, but further, by a secular re-
arrangement of molecules.   Thus two pieces of brass
which are of exactly equal length at a given date may
differ appreciably in length after a lapse of one hundred
years.

· While this defect is recognized, no satisfactory means
of avoiding its results has yet been discovered.

FIRST EXERCISE.—ALUMINIUM.

Data.*

		Molecular Weight.	Per Cent.
Al$_2$	27.0090 × 2	54.0	5.970
O$_3$	15.9633 × 3	47.9	5.296
(SO$_3$)$_3$	79.8739 × 3	239.6	26.490
(NH$_4$)$_2$	18.0210 × 2	36.0	3.980
O	15.9633	16.0	1.769
SO$_3$	79.8739	79.9	8.834
24 H$_2$O	(2 + 15.9633) 24	431.1	47.662
		904.5	100.001
Al$_2$	27.0090 × 2	54.0	52.993
O$_3$	15.9633 × 3	47.9	47.007
		101.9	100.000

The Compound Tested.

The substance tested is Ammonia alum, Ammonio-aluminic sulphate,

$$(NH_4)_2 SO_4 + Al_2 (SO_4)_3 + 24 H_2 O$$

The term alum is now applied to a class of salts, the members of the class being characterized by the following features:

(*a*) They tend to crystallize in cubes or in regular octahedrons.

(*b*) They have the chemical constitution represented by the general formula:

$$M^I_2 SO_4 + M^{IV}_2 (SO_4)_3 + 24 H_2 O.$$

In this formula, M^I_2 represents two atoms of a monad metal, and M^{IV}_2 represents two atoms of a tetrad (or pseudo-triad) metal.. It thus at once appears that the general formula is capable of representing a large num-

---

* For Table of Atomic Weights, see p. 182.

ber of alums.   The following examples illustrate a few of them :*

1. Potassio-aluminic alum, $K_2SO_4 + Al_2(SO_4)_3 + 24\ H_2O$
2. Sodio-aluminic alum, $Na_2SO_4 + Al_2(SO_4)_3 + 24\ H_2O$
3. Ammonio-aluminic alum, $(NH_4)_2SO_4 + Al_2(SO_4)_3 + 24\ H_2O$
4. Potassio-chromic alum, $K_2SO_4 + Cr_2(SO_4)_3 + 24\ H_2O$
5. Ammonio-ferric alum, $(NH_4)_2SO_4 + Fe_2(SO_4)_3 + 24\ H_2O$

Potassio-aluminic alum has long been the common alum of commerce ; of late, Ammonio-aluminic alum has taken

FIG. 16.—Mass of Alum Crystals.

its place.   The recent discovery of cheaper sources of Potassium, seems to have produced a tendency toward the renewed use of the potash alum.

Outline of the Process.

(*a*) Precipitate all the Aluminium as Aluminic hydroxide $(Al_2O_6H_6)$.

* Roscoe & Schorlemmer's Chemistry, Vol. II., Part I., pp. 51, 451.   Watts's Dictionary, Vol. V., pp. 578, 580, 588.

(*b*) After filtering, etc., change the Aluminic hydroxide to Aluminic oxide ($Al_2 O_3$), and weigh it in the latter form.

(*c*) Make the necessary calculations.

The Process.

The Weighing. — Weigh about one gramme of the pulverized Alum.

The Dissolving. — Dissolve the weighed salt in hot distilled water. Add a few drops of pure hydrochloric acid,[1] and some solution of ammonium chloride.[2]

The Precipitating. — To the solution, add ammonium hydroxide, cautiously, but in quantity sufficient to afford slight alkaline reaction.[3] Bring the solution to boiling, and maintain the boiling during two or three minutes.[3] Allow the precipitate to subside.

$$(N H_4)_2 S O_4 + Al_2 (S O_4)_3 + 6 N H_4 O H$$
$$= Al_2 O_6 H_6 + 4 (N H_4)_2 S O_4.$$

The Filtering. — Pour the clear liquid through a filter; then transfer the precipitate to the same filter, carefully washing it with boiling water.

The Burning. — Dry the precipitate thoroughly.[5] After drying, transfer the dried precipitate from the filter to a piece of glazed paper. Heat the filter-paper in a crucible until the tarry matters are burned off and the ash is white. Cool the crucible; then introduce the precipitate. Ignite the whole for some time.

$$Al_2 O_6 H_6 \text{ } heated = Al_2 O_3 + 3 H_2 O.$$

Place the hot crucible, and its contents, in a desiccator to cool. Finally weigh.

The Calculations. — From the weight of the contents of the crucible subtract the weight of the filter-ash. From the weight of the Alumnic oxide thus obtained, calculate the weight of Aluminium; calculate what per cent. this weight of Aluminium is of the amount of alum from which it was first obtained. Compare this percentage by test with the theoretical percentage; the difference is the error of the analytical operations.

Notes.

1. Hydrochloric acid is added for the purpose of affording a clear solution. If anything remains insoluble, it should be removed by filtration.

2. Ammonium chloride has the property of rendering Aluminic hydroxide insoluble in Ammonium hydroxide.

3. Of the three common alkalies—Potassium hydroxide, Sodium hydroxide, and Ammonium hydroxide—the latter is the only one suitable for use as the precipitant of Aluminium. For, while Potassium hydroxide and Sodium hydroxide afford, at first, precipitates of Aluminic hydroxide, these alkalies afterward redissolve the precipitate at first formed.*

The same precipitate is also *slightly* soluble in Ammonium hydroxide, hence a considerable excess of the latter should be avoided; but Ammonium salts diminish this solvent action of Ammonium hydroxide.

Boiling assists to expel the excess of Ammonium hydroxide, and so to favor complete precipitation of the Aluminic hydroxide.

4. Thorough washing of the precipitate is necessary; it tends strongly to retain some Ammonium sulphate. The last portions of this salt are removed by a final ignition with the blast-lamp.

If Ammonium chloride is not completely removed by the washing, it volatilizes during ignition; but under these circumstances, it carries off with it some Aluminic chloride that it forms at the high temperature. Of course this occasions loss.

5. If the precipitate is not thoroughly dried, it retains some water inside of its lumps; the steam formed, during heating, bursts the lumps, causing loss.

---

* Appleton's Qualitative Analysis, p. 34.

SECOND EXERCISE—ANTIMONY.

Data.

	Molecular Weight.	Per Cent.
K	39.	11.758
Sb	120.	36.177
O	16.	4.824
H_2	2.	0.603
O_4	63.9	19.264
$C_4H_2O_2$	81.8	24.664
$\frac{1}{2}(H_2O)$	9.	2.713
	331.7	100.003
Sb_2	239.9	71.42
S_3	96.	28.58
	335.9	100.00
Sb_2	239.9	78.966
O_4	63.9	21.034
	303.8	100.000

The Compound Tested.

The substance tested is Tartar emetic, Potassio-antimonylic tartrate [$K, SbO, H_2O_4, (C_4H_2O_2) + \frac{1}{2}H_2O$].

Tartaric acid has the formula, $\qquad H_4O_4(C_4H_2O_2)$.

It may be expanded to the form, $\begin{matrix} H_2 = \\ H_2 = \end{matrix} O_4 \begin{matrix} = \\ = \end{matrix} (C_4H_2O_2)$,

or it may be abridged to $H_4O_4\overline{T}$.

Two of the hydrogen atoms are replaceable by metals, two are not. Thus we may form such compounds as

$$\text{potassium tartrate,} \quad \begin{matrix} K_2 \\ H_2 \end{matrix} \Big\} O_4\overline{T},$$

$$\text{and hydro-potassium tartrate,} \quad \begin{matrix} H\ K \\ H_2 \end{matrix} \Big\} O_4\overline{T}.$$

Further, the compound radicle Antimonyl (SbO) may be substituted for one of the replaceable hydrogen atoms, so as to form such compounds as

Hydro-antimonylic tartrate, $\left.\begin{array}{l} H, Sb\ O \\ H_2 \end{array}\right\} O_4\ \overline{T},$

and Potassio-antimonylic tartrate, $\left.\begin{array}{l} K\ SbO \\ H_2 \end{array}\right\} O_4\ \overline{T}.$

Outline of the Process.

(*a*) Precipitate the Antimony as Sulphide $(Sb_2 S_3 + nS)$ by Sulphuretted hydrogen.

(*b*) Dry the Sulphide on a balanced filter and weigh it.

FIG. 17.—Apparatus for Precipitation of Antimony as Sulphide.

(*c*) The precipitate is usually contaminated with adhering Sulphur; hence corrections must be made. With this in view, divide the precipitate into two parts; in one part determine the Antimony as tetroxide $(Sb_2 O_4)$; in the other part determine the amount of Sulphur as Barium sulphate, and then by subtraction learn the amount of Antimony.

(*d*) Calculate the Antimony in three ways : first, from the weight of $Sb_2 S_3 + nS$; second, from the weight of $Sb_2 O_4$; third, from the weight of $Ba S O_4$.

The Process.

The Weighing. — Weigh about one gramme of Tartar-emetic.

The Dissolving. — Dissolve the weighed Tartar-emetic in water containing 10 c. c. of dilute Hydrochloric acid.[6] Add about one gramme of Tartaric acid.[7]

The Precipitating. — Into the solution pass a current of clean Sulphydric gas $(H_2 S)$[8] until the Antimony is entirely precipitated. (Prepare some fresh $H_2 S$ water for the subsequent washing of the precipitate.) Allow the precipitate to subside.

$$2\,(K, SbO, H_2 O_4 \overline{T}) + 4\,H_2 S = Sb_2 S_3 + K_2 S + 2\,H_4 O_4 \overline{T} + 2\,H_2 O.$$

The Filtering. — Pass the clear liquid through a balanced filter;[9] wash the precipitate two or three times with the sulphuretted hydrogen-water; finally transfer the precipitate without loss to the same filter. Pass the clear filtrates through the counterbalanced filter.

The Drying. — Dry the precipitate upon its filter – and also the other filter — for several hours at 212° F.[10] Then weigh the one filter against the other, adding the necessary weights. (Preserve the precipitate for subsequent analysis.)

The Calculation. — From the weight of the Sulphide (usually contaminated with adhering sulphur) make an *approximate* calculation of the amount of Antimony.

5*

Correction A.

The Weighing. — Carefully weigh one portion of the dry precipitate ($Sb_2 S_3 + nS$) into a porcelain crucible for Correction A. (Preserve another portion for Correction B.)

The Oxidizing. — To the portion of the precipitate in the porcelain crucible (for which a glass cover should be provided) add fuming Nitric acid,[11] drop by drop. A violent reaction ensues.

$$Sb_2 S_3 + S + 32 HNO_3 = Sb_2 O_4 + 4 H_2 SO_4 + 12 H_2 O + 16 N_2 O_4$$

When the reaction appears to be complete, evaporate the mixture to dryness. Then ignite strongly. The residue should be Antimony tetroxide ($Sb_2 O_4$). From its weight calculate the amount of Antimony.

Correction B.

The Weighing. — Weigh another portion of the dried Sulphide into a beaker for Correction B.

The Oxidizing. — To the weighed substance ($Sb_2 S_3 + nS$) add some fuming Nitric acid, drop by drop. (The acid must be free from Sulphuric acid.)[11] The Antimony compound is either wholly or partly oxidized. Evaporate the solution nearly to dryness. If any unoxidized Sulphur appears to be left, oxidize it by cautious addition of small fragments of Potassium chlorate.

$$Sb_2 S_3 + S + 32 HNO_3 = Sb_2 O_4 + 4 H_2 SO_4 + 12 H_2 O + 16 N_2 O_4.$$
$$S + 2 HNO_3 + 2 KClO_3 = K_2 SO_4 + 2 HNO_3 + Cl_2 + O_2.$$

When the Sulphur is completely oxidized, again evaporate nearly to dryness; dilute with water, boil and filter.

The Precipitating. — To the filtrate add a small quantity of pure Hydrochloric acid; boil the solution, and then determine the Sulphur as Barium sulphate.

Notes.

6. Excess of hydrochloric acid must be avoided, since the precipitated Sulphide of Antimony is somewhat soluble in it; this solubility is lessened by presence of abundance of Sulphuretted hydrogen.

7. Tartaric acid has a peculiarly favorable influence upon the solution of Antimony compounds; it prevents the precipitation by water.

8. Sulphuretted hydrogen is prepared by the action of dilute Sulphuric acid upon Ferrous sulphide.

$$Fe\,S + H_2\,SO_4 = Fe\,SO_4 + H_2\,S.$$

The Ferrous sulphide should be broken into lumps of about one-half an inch in diameter, and should be in relatively large quantity; a very large surface being thus afforded, a feeble action of the acid is able to give a steady flow of the gas for a long time. The Sulphuric acid should be dilute; otherwise the Ferrous sulphate produced by the reaction crystallizes; it thus forms a hard coating, which protects the Ferrous sulphide from further action of the acid, and so stops the evolution of the gas. After each period of use, the Ferrous sulphide should be thoroughly washed and then left to soak in water.

The gas should be cleaned by passage through a tube filled with loose cotton. (The gradual blackening of the cotton shows that it has detained matter which otherwise would have been carried along by the current of gas, to the injury of the solution.)

9. To prepare balanced filters, take two that have been kept in the same place, and try them upon the two pans of the balance. From the heavier, cut repeated small portions until its weight equals that of the other. During the process of the subsequent analysis, these two filters are subjected to the same washing influences, and the same drying influences, etc.; it is therefore assumed that when they are finally upon the balance with the precipitate upon one of them, the paper of the two filters

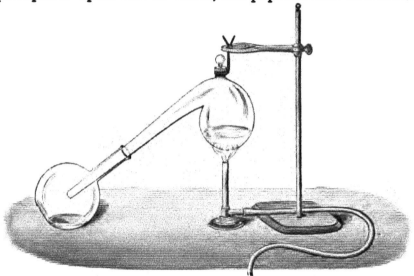

FIG. 18.—Apparatus for Preparation of fuming Nitric Acid.

will still exactly counterbalance each other, and that any increase of weight will be due solely to the precipitate.

10. The Antimonious sulphide, dried at 212° F., retains some water. It gives all off, however, after long drying, at 392° F.

11. Fuming Nitric acid may be prepared for this exercise. For this purpose, place in a litre retort

500 grammes Potassium nitrate

300 cubic centimetres Oil of vitriol.

Gently heat the mixture, and conduct the distillate into a flask which is cooled by snow, or ice-water, or ordinary running water.

$$KNO_3 + H_2SO_4 = HNO_3 + HKSO_4.$$
$$\underbrace{101 \qquad 98}_{199} \qquad \underbrace{63 \qquad 136}_{199}$$

About 250 cubic centimetres of Nitric acid may be obtained in a few hours. The acid should be carefully tested for Sulphuric acid, from which it must be free.

THIRD EXERCISE—ARSENIC.

(THE MAGNESIA METHOD.)

Data.

	Molecular Weight.	Per Cent.
As$_2$	149.8	75.771
O$_3$	47.9	24.229
	197.7	100.000
NH$_4$	18.0	4.541
Mg	24.0	6.054
O$_3$	47.9	12.084
AsO	90.9	22.931
12 H$_2$O	215.6	54.390
	396.4	100.000
Mg$_2$	48.0	15.509
As$_2$	149.8	48.401
O$_7$	111.7	36.090
	309.5	100.000

The Compound Tested.

The substance tested is Arsenious oxide, called White arsenic (As$_2$O$_3$). This substance tends to change into Arsenic oxide (As$_2$O$_5$) by absorption of oxygen. It is not easily wetted, so that small fragments of it frequently float upon water without being moistened. It dissolves in alkaline solvents, forming Arsenites, thus:

$$As_2O_3 + 6\,NaOH = 2\,Na_3AsO_3 + 3\,H_2O.$$

Outline of the Process.

(*a*) Oxidize the compound (if necessary) so as to change it from the Arsenious to the Arsenic form.

(*b*) Precipitate the Arsenic as Ammonio-magnesium arsenate ($NH_4MgAsO_4 + 12 H_2O$).

(*c*) Change the dried precipitate to Magnesium pyro-arsenate ($Mg_2 As_2 O_7$); from the weight of the latter, calculate the weight of the Arsenic.

The Process.

The Weighing.—Weigh about one gramme of the White arsenic.

The Dissolving.—Gently warm the weighed substance for some time with concentrated hydrochloric acid.[12]

$$As_2O_3 + 6 HCl = 2 As Cl_3 + 3 H_2O.$$

To this solution, when prepared, cautiously add a few small fragments of Potassium chlorate; the chlorate oxidizes the Arsenious compound.

$$As_2O_3 + 4 K Cl O_3 + 14 H Cl = 2 As Cl_5 + 4 K Cl + 7 H_2O + 2 Cl_2O_4.$$

The chlorate should be added until the solution acquires a decided smell of chlorine compounds. After the oxidation, place the solution in a warm place and under a ventilating-hood, and allow it to stand for several hours, in order that the Oxides of chlorine may escape.

The Precipitation.— To the solution, add a measured and calculated amount of previously prepared Magnesia solution,[13] and then sufficient Ammonium hydroxide to impart alkaline reaction to the entire liquid. Now allow the whole to remain at rest for from twelve to twenty-four hours.

$$As Cl_5 + Mg Cl_2 + 8 NH_4 OH + 8 H_2O$$
$$= (NH_4 Mg As O_4 + 12 H_2O) + 7 NH_4 Cl.$$

The Filtering.— Pass the clear liquid through a filter; then transfer the principal portion of the precipitate to the same filter, without washing. (Reserve the unwashed beaker.) Redissolve the precipitate by passing dilute hydrochloric acid through the filter, and into the reserved beaker; afterward reprecipitate by adding to the filtrate some Ammonia and a small amount of Magnesia-solution.[14] Again allow from twelve to twenty-four hours' time for the precipitate to form. At length, filter upon a balanced filter, but avoid washing the precipitate excessively,* as it is not very insoluble.[15]

The Drying.—Dry the precipitate at 212° F., and weigh it as $NH_4\, Mg\, As\, O_4 + \frac{1}{2}\, H_2\, O$.

Correction for this Method.

Carefully remove the Ammonio-magnesium arsenate from the filter. Saturate the filter-paper with a strong solution of Ammonium nitrate ; then dry it and carefully burn it in a weighed porcelain crucible.[16] Now transfer the reserved precipitate to the same crucible, and dry the whole for three hours at 260° F., to expel the Ammonium. Next, place the crucible over a lamp, heating gently at first, and afterwards heating very strongly. Finally raise the heat to redness and maintain it thus for one hour. Weigh the residue as Magnesium pyro-arsenate, $Mg_2\, As_2\, O_7$.

Notes on the Magnesia Method.

12. Solutions of Arsenic in hydrochloric acid may suffer a loss of volatile Arsenious chloride when an attempt is made to concentrate them by evaporation. When concentration is necessary, it should be preceded by addition of Ammonium hydroxide.

13. Prepare Magnesia-solution as follows :

Weigh, 11 grammes crystallized Magnesium chloride ($Mg\, Cl_2.6\,H_2O$),
    28   "   crystallized Ammonium chloride ($NH_4\, Cl$),
Measure, 70 cubic centimetres solution of Ammonium hydroxide (.960).
Add sufficient water to produce 200 c. c. of the solution.

---

* In the Journal of the Chemical Society, London, August, 1877, alcohol is recommended for this washing.

Allow the whole to stand several days. It is well to filter the solution before use. One cubic centimetre of this solution should be sufficient to precipitate 13 milligrammes of $As_2 O_3$ (or 16 mg. of $As_2 O_5$, or 10 mg. of $P_2 O_5$).

Test the solution as follows: measure 20 c. c. of it, then add 20 c. c. ammonium hydroxide (sp. gr. .960), then allow the mixture to rest twenty-four hours. No precipitate should be produced.

14. The redissolving and reprecipitation are to prevent the formation of Magnesium arsenate instead of Ammonio-magnesium arsenate.

15. The Magnesia precipitate, being somewhat soluble, is partly lost in the filtrate: this loss may be allowed for by adding to the weight first observed, $\frac{1}{30}$ milligramme for each cubic centimetre of the filtrate (not of the wash-water).

16. The purpose of the Ammonium nitrate is to oxidize the carbon and hydrogen of the filter-paper and to prevent their exercising a reducing action (and hence a tendency to volatilization) upon the Arsenic compound.

FOURTH EXERCISE—ARSENIC.
(THE SULPHIDE METHOD.)

Data.

	Molecular Weight.	Per Cent.
As$_2$	149.8	75.771
O$_3$	47.9	24.229
	197.7	100.000
As$_2$	149.8	60.944
S$_3$	96.	39.056
	245.8	100.000

The Compound Tested.

The substance tested is Arsenious oxide, called White arsenic (As$_2$ O$_3$). The substance tends to change to Arsenic oxide (As$_2$ O$_5$) by absorption of oxygen. It is not easily wetted, so that small fragments of it frequently float upon water without being moistened. It dissolves in alkaline solvents to form arsenites, thus:

$$As_2\ O_3 + 6\ Na\ OH = 2\ Na_3\ As\ O_3 + 3\ H_2\ O.$$

Outline of the Process.

(*a*) Precipitate the arsenic as sulphide (As$_2$ S$_3$ + nS) and weigh it, as such, on a balanced filter.

(*b*) Wash the precipitate with Carbon di-sulphide to remove the excess of sulphur; then weigh again.

The Process.

The Weighing.—Weigh about one gramme of White arsenic.

6

The Dissolving. — Dissolve the weighed substance by gently warming it for some time in concentrated hydro-chloric acid. (See page 58.) Dilute the solution with a considerable amount of water.

The Precipitating. — Into the solution, pass a current of clean Sulphuretted-hydrogen gas, and continue the flow of gas until the precipitation is complete.[8] The precipitate should readily subside in flakes; the *solution* should retain the odor of the gas. (Prepare some Sulphuretted-hydrogen water for the subsequent washing.)

$$2 \, As \, Cl_3 + 3 \, H_2 \, S = As_2 \, S_3 + 6 \, H \, Cl.$$

If the solution of arsenic contains that element in the *ic* form, precipitation by Sulphuretted hydrogen is tedious. The $H_2 S$ first reduces the Arsenic compound to the *ous* form, with evolution of sulphur; finally, $As_2 S_3$ is precipitated, but it is mixed with a variable amount of free sulphur.

The Filtering. — Transfer the precipitate to a balanced filter,[10] and wash it with $H_2 S$ water. Pass the washings through the second filter.

The Drying, etc. — Dry the two filters thoroughly at 212° F. Weigh the dried precipitate. Calculate the amount of arsenic, considering the precipitate to be pure $As_2 S_3$. The results are likely to be too high, owing to presence of free sulphur.

Correction for this Method.

Transfer either the whole or a weighed portion of the yellow precipitate to a small beaker. Then add Carbon di-sulphide to it. After a few minutes, filter the whole through a balanced filter. The $C S_2$ dissolves the free sulphur and leaves the $As_2 S_3$. Dry the latter at 212°. Weigh the purified precipitate and thence calculate the amount of arsenic.

FIFTH EXERCISE—BARIUM.

Data.

	Molecular Weight.	Per Cent.
Ba	136.8	56.204
Cl_2	70.7	29.047
$2 H_2O$	35.9	14.749
	243.4	100.000
Ba	136.8	58.788
O	16.	6.876
S	32.	13.752
O_3	47.9	20.584
	232.7	100.000

The Compound Tested.

The compound analyzed is crystallized Barium chloride $(Ba\ Cl_2 + 2 H_2O)$.

Outline of the Process.

Precipitate the barium as Barium sulphate $(Ba\ SO_4)$ in presence of *dilute* pure hydrochloric acid.

Weigh the Ba SO_4; from this weight calculate the amount of barium.

The Process.

The Weighing. — Weigh about one gramme of crystallized Barium chloride.

The Dissolving. — Dissolve the salt in water containing a few drops of pure hydrochloric acid.[18] Boil.

The Precipitating.—Add pure dilute Sulphuric acid drop by drop.[19] Boil the whole for some time.[20] (Have

boiling distilled water ready for the subsequent washing.)
Allow the precipitate to subside.

$$\text{Ba Cl}_2 + \text{H}_2 \text{SO}_3 = \text{Ba SO}_4 + 2 \text{ H Cl.}$$

The Filtering.—Decant the *clear* liquid upon a filter.
Wash the precipitate three or four times with boiling
water.[21]

The Burning.—Dry the precipitate; then transfer as
much as practicable from the filter to a square of glazed
paper; cover the precipitate with a clean beaker or bell-
glass.

Burn the paper to whiteness in a clean crucible. *Allow
the ash to cool;* then add to it one or two drops of pure
dilute Sulphuric acid.[22]

$$\text{Ba S} + \text{H}_2 \text{SO}_4 = \text{Ba SO}_4 + \text{H}_2 \text{S.}$$

Evaporate this acid to dryness, carefully avoiding the
loss by spattering, due to too great heat. Allow the crucible to cool again. Next transfer the rest of the precipitate — without loss — from the glazed paper to the crucible. Heat the whole once more, and when all volatile
matter is expelled transfer the crucible to a desiccator to
cool.

The substance weighed is Barium sulphate (Ba SO$_4$)
with the filter-ash.

Notes.

18. Hydrochloric acid is added for the purpose of holding in solution other compounds than Ba SO$_4$; but an
excess must be avoided, since even Ba SO$_4$ is somewhat
soluble in boiling concentrated H Cl (and in boiling concentrated H N O$_3$).

19. Ordinary Sulphuric acid often contains Lead sulphate (Pb SO$_4$); if such acid were used, it would give
rise to an error by reason of the fact that Pb SO$_4$ would

be thrown down with the Ba SO_4, and so give too high results.

20. If the solution is not boiled, the precipitated Ba SO_4 is so exceedingly fine that some of it on filtering passes through the paper. Boiling tends to make the precipitate collect into larger granules.

21. Barium sulphate has the property of carrying down with itself many other salts, even ordinarily soluble ones, that happen to be present with it in the solution. Even by thorough washing it is sometimes impossible to remove these salts completely. Barium chloride is a well-marked example of such a salt.

22. Filter-paper may be viewed as composed of nearly pure woody fibre, also called Cellulose ($C_6 H_{10} O_5$). When such paper is burned in contact with Ba SO_4, the latter is partly or wholly decomposed to Barium sulphide (Ba S).

$$Ba\ SO_4 + C_6\ H_{10}\ O_5 = Ba\ S + 5\ H_2\ O + 2\ CO_2 + 4\ C.$$

Subsequent addition of $H_2\ SO_4$ changes Ba S back to the sulphate.

$$Ba\ S + H_2\ SO_4 = Ba\ SO_4 + H_2\ S.$$

6 *   E

SIXTH EXERCISE—BISMUTH.

Data.

	Molecular Weight.	Per Cent.
Bi	207.5	72.705
O	16.	5.606
NO_3	61.9	21.689
	285.4	100.000
Bi_2	415.	89.652
O_3	47.9	10.348
	462.9	100.000

The Compound Tested.

The compound analyzed is Bismuthyl nitrate $(BiO NO_3)$, known in commerce as Basic nitrate of bismuth. It is insoluble in pure water, but is readily soluble in water containing a small quantity of hydrochloric or of Nitric acid.

Outline of the Process.

(*a*) Precipitate the Bismuth as sulphide.

(*b*) Dissolve the Bismuthous sulphide, and then reprecipitate it as a carbonate.

(*c*) Heat the dried carbonate so as to decompose it to Bismuthous oxide $(Bi_2 O_3)$, the substance which is to be weighed.

The Process.

The Weighing.—Weigh about one gramme of the Bismuthyl nitrate.

The Dissolving.—Dissolve this portion in water to which a few drops of hydrochloric acid are added.[23] Then dilute considerably with water—no matter if precipitation does occur.

The Precipitating, etc.—Pass clean Sulphuretted hydrogen gas into the solution, until precipitation is complete.[23 and 24]

$$2\,BiONO_3 + 3\,H_2S = Bi_2S_3 + 2\,HNO_3 + 2\,H_2O.$$

Filter, and wash with water containing H_2S.

Wash the moist Sulphide into a beaker, and redissolve it in *dilute* Nitric acid gently heated.[25]

$$Bi_2S_3 + 6\,HNO_3 = 2\,Bi(NO_3)_3 + 3\,H_2S.$$

Filter the solution; boil it to expel H_2S. Dilute the clear liquid if necessary. (Precipitation of the basic Nitrate does no harm.) Add a slight excess of Ammonium carbonate. Heat the solution nearly to the boiling point, and keep it at this temperature for some time.[27] Then allow the precipitate to subside.

$$2\,Bi(NO_3)_3 + 3\,(NH_4)_2CO_3 = Bi_2(CO_3)_3 + 6\,NH_4NO_3.$$

The Final Filtering.—Pass the clear liquid through a filter; wash the precipitate with water, and finally transfer it to the same filter.

The Burning.—Dry the precipitate; then remove it to a piece of glazed paper. Burn the filter-paper in a porcelain crucible. To the filter-ash, add a single drop of Nitric acid, evaporate carefully to dryness, and gently ignite.[27]

Cool the crucible, then introduce the precipitate, and ignite strongly for some time. Weigh as Bismuthous oxide (Bi_2O_3).

$$Bi_2(CO_3)_3 \; heated = Bi_2O_3 + 3\,CO_2.$$

Notes.

23. A large excess of hydrochloric acid interferes with the precipitation by Sulphuretted hydrogen, especially if the solution containing the bismuth is con-centrated.

24. The ultimate product ($Bi_2 O_3$) may be readily pro-duced by heating not only $Bi_2 (CO_3)_3$, but also Basic ni-trate. But it cannot be obtained by heating Bismuthyl chloride (Bi OCl) nor Bismuthyl sulphate (Bi O)$_2$ SO$_4$. Hence, solutions to be tested must be free from acids and acid radicles other than Nitric, since such solutions may produce compounds that will not yield $Bi_2 O_3$. This danger may be averted by previous precipitation of the bismuth as sulphide—with subsequent change to nitrate.

25. Care must be taken to dissolve the Bismuthous sulphide in *dilute* Nitric acid. When strong acid is used, it sometimes changes the Bismuthous sulphide—partly, at least—into a sulphate; this latter compound, upon the addition of Ammonium carbonate, changes to a basic Sulphate, which is precipitated in the place of the car-bonate desired.[24]

26. In *cold* solutions, Ammonium carbonate does not precipitate bismuth completely; hence the solutions must be heated.

27. The carbon and hydrogen of the filter-paper reduce oxide of bismuth to metallic Bismuth; hence the precip-itate must be carefully separated from the paper, before ignition.[22] By addition of Nitric acid, any small glob-ules of metallic bismuth adhering to the filter-ash are turned back to nitrate, and—by subsequent heating—to oxide.

SEVENTH EXERCISE—BROMINE.

Data.

	Molecular Weight.	Per Cent.
K	39.	32.828
Br	79.8	67.172
	118.8	100.000
Ag	107.7	57.44
Br	79.8	42.56
	187.5	100.00

The Compound Tested.

This is Potassium bromide (K Br). It should be free from Bromates, also from Chlorides and Iodides.

Outline of the Process.

Precipitate the Bromine as Silver bromide (Ag Br), and weigh as such. (Perform the process — as far as convenient — away from exposure to sunlight.[29])

The Process.

The Weighing. — Weigh about one gramme of the Potassium bromide.

The Dissolving. — Dissolve the weighed salt in water.

The Precipitating. — From the weight of the bromide used, calculate how much Silver nitrate will be needed for complete precipitation. Add a little more than the required amount of Silver nitrate solution, and slightly acidify with $H N O_3$.[28]

$$K Br + Ag NO_3 = Ag Br + K N O_3.$$

During addition of Silver solution, gradually warm the

mixture,[28] and actively stir it; both processes help the precipitate to collect. When the full amount of Silver solution (or a slight excess) has been added, the precipitate gathers into masses, and the solution becomes clear.

The Filtering. — Filter the precipitated Silver bromide and wash it.

The Burning. — Dry the precipitate and separate it from the paper; burn the paper first. When the crucible is cool, add a drop of Nitric acid; warm, and then add a drop of hydrobromic acid and evaporate to dryness.

$$2\,Ag\,Br + C_6\,H_{10}\,O_5 = 2\,Ag + 2\,H\,Br + 4\,H_2\,O + CO + 5\,C$$
$$(CO + 5\,C) + 5\,O_2 + O = 6\,CO_2.$$
$$6\,Ag + 8\,H\,N\,O_3 = 6\,Ag\,NO_3 + N_2\,O_2 + 4\,H_2\,O.$$
$$Ag\,NO_3 + H\,Br = Ag\,Br + H\,NO_3.$$

Add the precipitate and heat it to incipient fusion; then cool, and weigh as Ag Br.[30]

Notes.

28. Solutions containing Bromides *and free acid* must not be boiled, for then there might be loss of Hydrobromic acid (H Br). But when Silver nitrate is added, the presence of Nitric acid is desirable; the latter both dissolves other substances present and assists the Silver bromide to collect.

29. Sunlight blackens Silver bromide with decomposition of the salt and loss of Bromine.

30. The Silver bromide adhering to the crucible may be removed by adding to it Hydrochloric acid and metallic Zinc. The nascent hydrogen thus liberated decomposes the Silver bromide, leaving metallic Silver in a condition such that it may be readily removed.

$$2\,Ag\,Br + Zn + 2\,H\,Cl = Zn\,Cl_2 + 2\,H\,Br + 2\,Ag.$$

EIGHTH EXERCISE— CALCIUM.

Data.

	Molecular Weight.	Per Cent.
Ca	40.	40.040
O	16.	16.016
CO$_2$	43.9	43.944
	99.9	100.000
Ca	40.	71.429
O	16.	28.571
	56.	100.000

The Compound Tested.

The compound examined is Calcium carbonate (Ca CO$_3$);

FIG. 19.—Illustration of Double Refraction by Iceland Spar.

the particular form of that substance selected is the clear mineral called Iceland spar, also double refracting spar.

Outline of the Process.

(*a*) Precipitate the Calcium as Calcium oxalate (CaO$_2$C$_2$O$_2$).

(*b*) Heat this precipitate until it decomposes, first into Calcium carbonate (Ca CO$_3$), and finally to Calcium oxide or Quick-lime (Ca O). Weigh the latter.

The Process.

The Weighing. — Reduce a small crystal of Iceland spar to a fine powder. Then weigh about one gramme of the powder.

The Dissolving. — Dissolve the Carbonate in a covered beaker, in pure dilute hydrochloric acid. Take care to avoid loss by effervescence. Boil the solution for a few minutes to expel Carbon dioxide (CO_2).[31]

$$Ca\ CO_3 + 2\ H\ Cl = Ca\ Cl_2 + CO_2 + H_2\ O.$$

The Precipitation. — To the solution add Ammonium hydroxide to alkaline reaction; then add a slight excess of a fresh solution of Ammonium oxalate $(NH_4)_2\ O_2\ (C_2\ O_2)$.[32] Allow the precipitate to subside.

$$Ca\ Cl_2 + (NH_4)_2\ O_2\ (C_2\ O_2) = Ca\ O_2\ (C_2\ O_2) + 2\ (NH_4)\ Cl.$$

The Filtration. — Pass the clear liquid through the filter. Wash by decantation, adding to each new portion of wash-water small quantities of Ammonium hydroxide and of Ammonium oxalate. These diminish the solubility of the Calcium oxalate. Finally transfer the precipitate to the filter.

The Burning. — Dry the precipitate thoroughly. Transfer paper and precipitate together to a deep platinum crucible. (For care of platinum vessels, see page 74.) Heat *gently* over a Bunsen lamp until the paper is burned white.[33] The Calcium oxalate changes to Calcium carbonate, with escape of Carbon monoxide.

Next heat the precipitate strongly with the blast-lamp for five minutes. This slowly changes the Calcium carbonate to Calcium oxide with evolution of Carbon dioxide.

$$Ca\ O_2\ C_2\ O, heated = Ca\ CO_3 + CO.$$
$$Ca\ CO_3\ heated = Ca\ O + CO_2.$$

The Weighings. — Cool the crucible and contents in

a desiccator and weigh.[33] Now heat it a second time over the blast-lamp and again weigh.[33] If the weight indicates that the first heating had not expelled all of the Carbon dioxide, heat yet a third time. In general, repeat the heating until two successive weighings give identical results. The residue should be solely Calcium oxide and filter ash. (After the last weighing, it is well to put the Calcium oxide into a small clean beaker with a few drops of water, and then to add a drop of dilute Hydrochloric acid. The Calcium oxide should dissolve, *without effervescence*, to a clear solution.)

Notes.

31. If Carbon dioxide be not expelled, that portion remaining in the solution combines with Ammonium hydroxide to form Ammonium carbonate.

$$2 \, (NH_4) \, OH + CO_2 = (NH_4)_2 \, CO_3 + H_2 O.$$

This afterward forms a white precipitate of Calcium carbonate, which mixes with the Calcium oxalate. But as Calcium carbonate is more soluble than Calcium oxalate, it is well to avoid the formation of the former.

32. Ammonium oxalate, in solution, undergoes a slow decomposition, forming, among other things, Ammonium carbonate. For quantitative analysis, therefore, it is preferable to make up a fresh solution from the crystalline oxalate. To do this, calculate the amount of Ammonium oxalate demanded by the supposed amount of Calcium to be precipitated; then weigh roughly a somewhat larger quantity and dissolve it in water, filtering, if necessary, before use.

33. Filter-papers burn at a gentle heat better than at a very strong one; the latter seems to liberate from the cellulose or woody fibre ($C_6 H_{10} O_5$) of the paper, *a diffi-*

7

cultly combustible carbon resembling graphite. The first
heating changes a portion of the Calcium oxalate to
Calcium oxide, so that it is not safe to weigh it as
Ca CO$_2$. Of course, at a later stage a very high temperature is necessary to change the whole of the Calcium
carbonate to Calcium oxide.

34. Exposure of Calcium oxide to the air before weighing should be avoided. It is liable to gain largely in
weight by absorption of both Carbon dioxide and water.

35. Great care must be exercised in the use of platinum vessels. When heated, they should rest on platinum triangles. There should not be heated in them
matters containing, nor likely to afford, any of the following substances :

Chlorine,

Potassium nitrate,

Potassium hydroxide,

Metals, or sulphides of them,

Easily deoxidizable metallic oxides,

Organic metallic salts,

Phosphates in presence of organic matters.

If platinum vessels become soiled, they may be gently
rubbed with round sand. If necessary, a little Hydro-
potassium sulphate, or a little Borax may be fused in them.

NINTH EXERCISE—CARBON DIOXIDE.
(THE IGNITION METHOD.)

Data.

	Molecular Weight.	Per Cent.
Ca	40.	40.040
O	16.	16.016
CO_2	43.9	43.944
	99.9	100.000
C	12.	27.335
O_2	31.9	72.665
	43.9	100.000

The Compound Tested.

The compound used is Calcium carbonate $(Ca\ CO_3)$; the particular form of that substance selected is the clear mineral called Iceland spar, also double refracting spar.

Outline of the Process.

Heat the substance to a high temperature, and weigh it after the gas (CO_2) has been expelled.

The Process.

The Weighing. — Weigh about 500 milligrammes of powdered Iceland spar.

The Heating. — Place the powder in a deep platinum crucible, and heat it with the blast-lamp for about five minutes. Cool in the desiccator and weigh. Repeat the heating, and weigh again. If the second heating does not occasion additional loss of weight, it may be assumed that the whole of the Carbon dioxide is expelled. If it does occasion loss, continue the heating until the weight does not change. (See pp. 72 and 73.)

Calculate the loss of weight as Carbon dioxide.

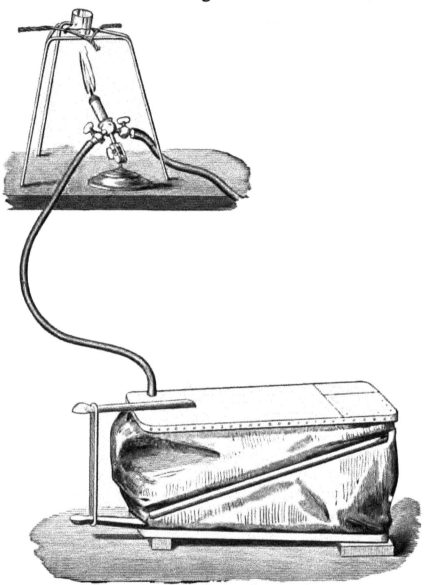

FIG. 20.—Apparatus for Heating Calcic Carbonate.

Note.

36. Of course this process is not applicable to sub-stances containing either moisture or **any** volatile con-stituents. It is **also** inapplicable to certain carbonates that do not lose their Carbon dioxide by heating.

TENTH EXERCISE—CARBON DIOXIDE.
(JOHNSON'S METHOD.)

Data.

	Molecular Weight.	Per Cent.
Ca	40.	40.040
O	16.	16.016
CO_2	43.9	43.944
	99.9	100.000
C	12.	27.335
O_2	31.9	72.665
	43.9	100.000

The Compound Tested.

The compound tested is calcium carbonate (Ca CO_3); the particular form of that substance selected is the clear mineral called Iceland spar, also double refracting spar.

Outline of the Process.

Expel the Carbon dioxide from a weighed amount of carbonate by some liquid acid, the operation being conducted by means of an apparatus made for the purpose. After the gas is expelled, note the loss of weight of the apparatus.

The Apparatus.

Numerous forms of apparatus have been suggested; the special one recommended is that devised by Professor Samuel W. Johnson. (See Fig. 21.)

A, B, C, contains dilute hydrochloric acid (specific gravity 1.1).[37]

7 *

FIG. 21.—Johnson's Apparatus for Determination of Carbon Dioxide.

D, is an exit tube, containing cotton below and Calcium chloride[38] above.

F, is the bottle in which is placed the carbonate to be analyzed.

G, is a generator of carbon dioxide; it has a drying-tube.

The Process.

Prepare the apparatus in accordance with the figure.

First Stage. Weigh about 1 gramme of Calcium carbonate in fragments, and transfer it dry to F. Connect the stoppers at C and D, and adjust all the tubes in place as in the cut. Let a steady but gentle stream of Carbon dioxide from a generator pass through the whole apparatus for about fifteen minutes.[39] Disconnect the generator at A, and stop the tubes A and E. Weigh the entire bulb-apparatus, A, B, C, D, E, F.

Second Stage. Gently incline the apparatus so that a few drops of acid will flow from B to F. Continue to transfer this acid according to requirement; let it act until the carbonate in F has completely dissolved.

Third Stage. Again connect the bulb-apparatus at A with the Carbon dioxide generator, and pass dried Carbon dioxide through the apparatus for a minute.

Fourth Stage. Finally replace the stoppers at A and E, and weigh the bulb-apparatus again.

The loss of weight indicates the amount of Carbon dioxide expelled from the carbonate tested.

Notes.

37. Hydrochloric acid is selected as the expelling acid, because it forms easily-soluble Calcium chloride. If Sulphuric acid is used it forms an insoluble and pasty Calcium sulphate (plaster of Paris), which (*a*) coats over a portion of the Calcium carbonate, and so prevents its complete decomposition ; (*b*) entangles Carbon dioxide, which it is desired to expel.

38. Calcium chloride is used in the drying tubes because of its great attraction for water. Cotton retains the larger drops of water ; thus the efficiency of the Calcium chloride is maintained.

39. Carbon dioxide is prepared in the generator by Calcium carbonate and Hydrochloric acid. It is best to use a very large quantity of small fragments of Calcium carbonate ; a very gentle action of Hydrochloric acid over a large surface gives a slow and constant current of Carbon dioxide. Of course this gas is dried by the drying-tube.

The purpose of the current of Carbon dioxide is best understood by considering the effects of neglecting to use it :

First. A part of the liberated gas would be held in solution by the Hydrochloric acid used. (By Johnson's method this acid is previously saturated with Carbon dioxide ; hence it cannot take any more.)

Secondly. A part of the liberated gas would stay in the bulb-apparatus, having expelled its volume of air. This would give low results, as illustrated by the following example :

Suppose the bulb-apparatus to have a volume of only 50 cubic centimetres. Then the final weight of the apparatus may vary 32 milligrammes according as its space is filled with air or with Carbon dioxide. (For CO_2 weighs 1.86 milligrammes per cubic centimetre and air weighs 1.22 milligrammes per cubic centimetre, both being at 60° Fahrenheit. Thus there is a difference of weight of .64 of a milligramme per cubic centimetre. Hence in 50 cubic centimetres there may be a difference equal to .64 × 50 = 32 milligrammes.) If, then, 1 gramme of the Carbonate is used for the test, an error of as great as three per cent. may readily result from the mere uncorrected displacement of air by the Carbon dioxide.

Thirdly. Some samples of Calcium chloride contain Calcium hydrate ; such samples when used for the first time in the exit tube will retain a portion of the Carbon dioxide, and so give rise to a low final result. By use of the current of CO_2 such Calcium hydrate will be neutralized, and give no further trouble.

40. The bulb-apparatus should be handled with a thick towel or with wooden tweezers, since the heat of the hands may expand and expel Carbon dioxide.

NOTE.—This method is not applicable to carbonates of the alkali metals, for they absorb an extra molecule of Carbon dioxide during the first stage of the process.

ELEVENTH EXERCISE—CARBON DIOXIDE.
(SCHEIBLER'S METHOD.)

Data.

	Molecular Weight.	Per Cent.
Ca	40.	40.040
O	16.	16.016
CO_2	43.9	43.944
	99.9	100.000
C	12.	27.335
O_2	31.9	72.665
	43.9	100.000

The Compound Tested.

The compound examined is Calcium carbonate ($Ca\ CO_3$), the particular form of that substance selected being the clear mineral called Iceland spar, also double refracting spar.

Outline of the Process.

The process is that devised by Dr. Scheibler.

Expel the Carbon dioxide from its compound and collect the gas expelled; determine the bulk of the gas by the volume of water it displaces from a measuring-tube. Now, from the known weight of one cubic centimetre of Carbon dioxide, estimate the weight of the volume of gas obtained.

The analysis is conducted by help of an apparatus devised for the purpose.

F

Scheibler's Apparatus.

The principal parts of the apparatus are seven:

First. A test-bottle A, containing an acid-tube S, and having a connecting-tube *r;* to this tube is attached a

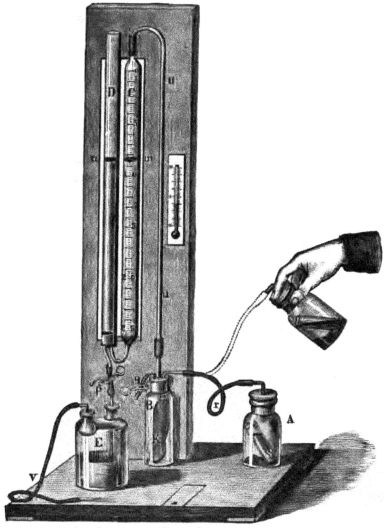

FIG. 22.—Scheibler's Apparatus for Determination of Carbon Dioxide.

rubber balloon K, which is to hold the Carbon dioxide liberated from the test-bottle A.

Second. An air-bottle B. Its tubes *q, r, u,* must pass air-tight through the cork.

Third. A measuring-tube C. This must be graduated from zero at the top to about 150 cubic centimetres at the bottom, and graduated in one-half cubic centimetres.

Fourth. A water-tube D, always open at the top. At the bottom it connects with the measuring-tube by a tube that is also always open. It also connects below with the bottom of the water-bottle by a tube having a tap at P.

Fifth. A water-bottle E, constructed like a wash-bottle. By blowing at V, and opening the tap P, water may be forced up into the water-tube D to any desired height.

Sixth. A Thermometer.

Seventh. A Barometer.

The Process.

First Stage. Weigh about ½ of a gramme of finely pulverized Iceland spar.[41] Transfer it dry to bottle A. In the tube S place 5 cubic centimetres of hydrochloric acid of specific gravity 1.12. Adjust this tube, without spilling, in the test-bottle A.

Open the tap on the air-bottle at q, then by blowing at V force water up into the water-tube D, so that the water stands at zero or above in the measuring-tube C, and at or above the same level in the water-tube.[42] Now, by opening the tap at P, draw down the water so that it stands at zero in the measuring-tube and at the same level in the water-tube.

Fit the cork belonging to A tight to its place.

Close the tap at q.

Second Stage. Incline the test-bottle A, so that a little acid may flow upon the carbonate. The liberated gas expands the balloon, expels air[43] from the air-bottle B into the measuring-tube C, depressing the water level in it. Now, by opening the water-tap at P, carefully keep the level in the water-tube at the same level as in the measuring-tube.[44] Continue pouring the acid, little by little, upon the carbonate.

Third Stage. When the operation is over, carefully adjust the water level in the water-tube to that in the measuring-tube. Note the number of cubic centimetres registered in the measuring-tube; note the temperature and the height of the barometer.[45]

The Calculations.—In the *absorption-table* (p. 86) find how many cubic centimetres of Carbon dioxide are held dissolved by the hydrochloric acid used, and add these to the indication on the gas-tube.[46]

From the *weight-table* (p. 87) find the weight in milligrammes of one cubic centimetre of Carbon dioxide at the temperature and barometric pressure of the experiment.[48] Multiply this weight by the number of cubic centimetres found; the product is the weight of the Carbon dioxide expelled from the substance tested. From this weight the per cent. is easily estimated.

Notes.

41. Scheibler's apparatus is best adapted to substances containing but a small amount of Carbon dioxide. Such a substance is bone-black, for whose analysis for Carbon dioxide the process was devised. It is evident that not more than *half a gramme* of Calcium carbonate can be used as above described. Pure Calcium carbonate yields 44 per cent. of Carbon dioxide; from half a gramme of Iceland spar this is equal to 220 milligrammes of Carbon dioxide. Since Carbon dioxide weighs 1.8 milligrammes per cubic centimetre at ordinary temperature (60° F.) and pressure (29.92 inches), 220 milligrammes are equivalent to about

$\frac{220}{1.8} = 122$ cubic centimetres; but the capacity of the measuring-tube is 150 cubic centimetres.

42. If the water levels are not at the same height in both the water-tube and the measuring-tube, then the air in the portions of apparatus between C and K may be in a condition either of condensation or of rarefaction as compared with the outer atmosphere; either condition is inadmissible. Further, there is a tendency (according as one or the other level is higher) to leakage of air out of the apparatus or to leakage of air into the apparatus.

43. The rubber bag retains the liberated Carbon dioxide. If the gas were allowed to come in direct contact with the water in the measuring-tube, some of the gas would be dissolved, and thus the volume indicated would be too low.

44. If during the evolution of the gas the water were allowed to rise in the water-tube, its extra pressure would be constantly offering resistance to the free evolution of the Carbon dioxide; it would also tend to produce loss of gas by leakage.

45. When the final measuring of gas takes place, the two water levels being the same, the gas is subjected to the atmospheric pressure at that moment; this is indicated by the barometer.

If the operator has a *higher pressure* than he supposes, of course the gas is thereby compressed, and results that are too low are reported. If the operator subjects the gas to a *lower temperature* than he supposes, the same error is suffered.

Opposite conditions accidentally obtained give rise to errors in the opposite direction.

46. It appears that the hydrochloric acid used absorbs

8

more Carbon dioxide the greater the amount of gas liberated; hence an absorption-table has been prepared for assistance in making this correction.

Absorption-Table,

Showing in cubic centimetres the amounts of Carbon dioxide dissolved by 5 c. c. of H Cl (Sp. gr. 1.125) for the specified amounts of gas evolved, expressed in cubic centimetres.

C. C. Evolved.	C. C. Absorbed.	C. C. Evolved.	C. C. Absorbed.	C. C. Evolved.	C. C. Absorbed.	C. C. Evolved.	C. C. Absorbed.	C. C. Evolved.	C. C. Absorbed.
1.	1.85	21.	4.95	41.	5.24	61.	5.50	81.	5.78
2.	2.00	22.	4.96	42.	5.25	62.	5.51	82.	5.79
3.	2.16	23.	4.97	43.	5.26	63.	5.52	83.	5.80
4.	2.31	24.	4.98	44.	5.27	64.	5.54	84.	5.82
5.	2.47	25.	5.00	45.	5.28	65.	5.55	85.	5.83
6.	2.62	26.	5.03	46.	5.30	66.	5.57	86.	5.85
7.	2.78	27.	5.04	47.	5.31	67.	5.58	87.	5.86
8.	2.93	28.	5.06	48.	5.32	68.	5.59	88.	5.87
9.	3.09	29.	5.07	49.	5.34	69.	5.61	89.	5.89
10.	3.24	30.	5.09	50.	5.35	70.	5.62	90.	5.90
11.	3.40	31.	5.10	51.	5.36	71.	5.64	91.	5.92
12.	3.55	32.	5.11	52.	5.37	72.	5.65	92.	5.93
13.	3.71	33.	5.13	53.	5.38	73.	5.66	93.	5.94
14.	3.86	34.	5.14	54.	5.40	74.	5.68	94.	5.96
15.	4.02	35.	5.16	55.	5.41	75.	5.69	95.	5.97
16.	4.17	36.	5.17	56.	5.43	76.	5.71	96.	5.99
17.	4.33	37.	5.18	57.	5.44	77.	5.72	97.	6.00
18.	4.48	38.	5.20	58.	5.45	78.	5.73	98.	6.01
19.	4.64	39.	5.21	59.	5.47	79.	5.75	99.	6.03
20.	4.79	40.	5.23	60.	5.48	80.	5.76	100.	6.04

Weight-Table,

Expressing in milligrammes the weight of one cubic centimetre of Carbon dioxide at different Temperatures from 10° C. to 25° C., and at different barometric pressures from 750 millimetres to 770 millimetres of mercury.

	750	752	754	756	758	760	762	764	766	768	770
10°	1.849	1.854	1.859	1.864	1.869	1.874	1.879	1.884	1.889	1.894	1.899
11°	1.841	1.846	1.851	1.856	1.861	1.866	1.871	1.876	1.881	1.886	1.891
12°	1.833	1.838	1.843	1.848	1.853	1.858	1.863	1.868	1.873	1.878	1.883
13°	1.825	1.830	1.835	1.840	1.845	1.850	1.855	1.860	1.865	1.870	1.875
14°	1.817	1.822	1.827	1.832	1.837	1.842	1.847	1.852	1.856	1.861	1.866
15°	1.809	1.814	1.818	1.823	1.828	1.833	1.838	1.843	1.848	1.853	1.858
16°	1.800	1.805	1.810	1.815	1.820	1.825	1.830	1.835	1.839	1.844	1.849
17°	1.792	1.797	1.802	1.807	1.811	1.816	1.821	1.826	1.831	1.836	1.841
18°	1.784	1.788	1.793	1.798	1.803	1.808	1.813	1.818	1.822	1.827	1.832
19°	1.775	1.780	1.785	1.790	1.794	1.799	1.804	1.809	1.814	1.819	1.823
20°	1.766	1.771	1.776	1.781	1.786	1.791	1.795	1.800	1.805	1.810	1.815
21°	1.758	1.763	1.767	1.772	1.777	1.782	1.787	1.791	1.796	1.801	1.806
22°	1.749	1.754	1.759	1.763	1.768	1.773	1.778	1.783	1.787	1.792	1.797
23°	1.740	1.745	1.750	1.755	1.759	1.764	1.769	1.774	1.778	1.783	1.788
24°	1.731	1.736	1.741	1.746	1.750	1.755	1.760	1.765	1.769	1.774	1.779
25°	1.722	1.727	1.732	1.736	1.741	1.746	1.751	1.755	1.760	1.765	1.770
	750	752	754	756	758	760	762	764	766	768	770

TWELFTH EXERCISE—CHLORINE.

(GRAVIMETRIC METHOD.)

Data.

	Molecular Weight.	Per Cent.
Na	23.	39.384
Cl	35.4	60.616
	58.4	100.000
Ag	107.7	75.262
Cl	35.4	24.738
	143.1	100.000

The Compound Tested.

This is crystallized common salt (Na Cl).

Chlorine combines with electro-positive elements, such as hydrogen and the metals, to form chlorides. It is to this class of compounds that the processes here described are applicable; but other compounds of chlorine — if previously changed into the form of chlorides — may be tested by the processes described in this and the next following Exercise.

Outline of the Process.

To the solution of the chloride add a slight excess of Silver nitrate (Ag NO$_3$), in presence of Nitric acid.

Weigh the precipitate of Silver chloride (Ag Cl) formed. Direct sunlight, and even abundance of diffused sunlight, must be avoided.[47]

The Process.

The Weighing.—Weigh about one gramme of pure Sodium chloride (Na Cl)· Heat it for awhile in a covered crucible, so as to expel the water retained inside the crystals. Cool in a desiccator, and weigh again, so as to get the exact weight of the dry salt.

The Dissolving.—Dissolve the salt in warm (not boiling) water.[48]

The Precipitation.—Add a few drops of pure Nitric acid, and then a slight excess of a solution of Silver nitrate (Ag NO$_8$).[50]

Fig. 23.—Cooling a Crucible in a Desiccator.

Heat the solution, and stir it vigorously; the precipitate is thus made to collect as heavy, curdy lumps, leaving the supernatant liquid almost perfectly clear.[49]

$$\text{Na Cl} + \text{Ag NO}_3 = \text{Ag Cl} + \text{Na NO}_3.$$

The Filtering.—Filter; wash the precipitate until the filtrate no longer affords a test for Silver, upon addition of hydrochloric acid to a small portion of it.

The Burning.—Dry the precipitate, and remove it from the paper as completely as possible. Burn the paper in a porcelain crucible. When the crucible is cool, add a drop of Nitric acid to the ash, and after a few minutes add a drop of Hydrochloric acid; then evaporate carefully to dryness. Next introduce the principal mass of the Silver chloride (Ag Cl); heat gently until the precipitate commences to fuse. Cool and weigh.[51]

8 *

Notes.

47. Direct or diffused sunlight decomposes Silver chloride to a purple or black compound (perhaps $Ag_2 Cl$), the change occasioning a loss of chlorine. This decomposition is effected principally by the actinic (chemical) rays associated with the violet end of the solar spectrum ; hence yellow light is preferred for the room where the analysis is performed.

48. The solution should not be boiled at first, lest the Nitric acid expel some of the Hydrochloric acid it has liberated. After the chlorine has combined with the silver, this danger need not be feared.

49. The presence of Nitric acid prevents the precipitation of many salts, other than Silver chloride, that might in certain cases be present ; it also in some way helps the precipitated Silver chloride to collect.

50. The weight of pure Silver nitrate absolutely *needed* should be calculated by use of the following proportion :

Molecular weight of Na Cl.	:	Molecular weight of Ag NO₃.	: :	Gross weight of Na Cl taken.	:	Gross weight of Ag NO₃ required.

51. Upon burning the filter-paper, the hydrogen of the Hydrocarbons thus formed, invariably abstracts Chlorine from any Silver chloride present ; metallic silver is left.

$$Ag\ Cl + H = Ag + H\ Cl.$$

Upon addition of Nitric acid, this silver is dissolved.

$$6\ Ag + 8\ H\ NO_3 = 6\ Ag\ NO_3 + N_2O_2 + 4\ H_2\ O.$$

Upon subsequent addition of Hydrochloric acid, Silver chloride is formed in quantity equal to that previously decomposed.

$$Ag\ NO_3 + H\ Cl = Ag\ Cl + H\ NO_3$$

52. To remove Silver chloride from a porcelain crucible, add to it a strip of zinc and some dilute Sulphuric acid.

The hydrogen so afforded decomposes the Silver chloride, leaving the metallic silver in a spongy condition such that it is easily detached.

$$Zn + H_2 SO_4 = Zn SO_4 + H_2.$$
$$2 Ag Cl + H_2 = 2 Ag + 2 H Cl.$$

THIRTEENTH EXERCISE—CHLORINE.
(VOLUMETRIC METHOD.)

Data.

	Molecular Weight.	Per Cent.
Na	23.	39.384
Cl	35.4	60.616
	58.4	100.000

The Compound Tested.

The substance tested is commercial Sodium chloride (Na Cl). The method is applicable only to chlorine in the form of chlorides; even then the solution must be *neutral.*

This necessity results from the fact that both feebly acid and feebly alkaline solutions decompose the indicator used — red Silver chromate ($Ag_2 Cr O_4$).

If therefore a given solution is *acid*, it may be prepared for the test by first neutralizing with an excess of pure Calcium carbonate ($Ca CO_3$). *Alkaline* solutions should be first rendered feebly acid with pure Nitric acid; then they should be neutralized — as just described — with Calcium carbonate.

The process is not applicable to such compounds as Bleaching-powder,

$$Ca Cl_2 + Ca O_2 Cl_2.$$

It is not applicable to chlorates, except after decomposition to chlorides.

Outline of the Process.

Standardize a solution of Silver nitrate (Ag NO$_3$) by use of pure Sodium chloride (Na Cl), so as to learn the exact amount of Sodium chloride — and hence by calculation of chlorine — that each cubic centimetre of the silver solution equals.

Use the silver solution to determine the amount of chlorine—as chloride—in other solutions.

The Standard Silver Solution.

The Weighing. — Weigh about 4 grammes of pure crystallized Silver nitrate (Ag NO$_3$); dissolve it in about 400 cubic centimetres of distilled water; transfer the solution to a clean glass-stoppered bottle.[68] (Avoid exposing the solution to direct sunlight.)

Dry about 300 milligrammes of crystallized Sodium chloride (Na Cl) as explained under the gravimetric test for chlorine (p. 89). Then weigh accurately three portions of it — not necessarily of equal weights. Dissolve these portions in water, separately, in three small beakers, carefully noting the quantity of salt in each. To each solution, add a few drops of *yellow* Potassium chromate (K$_2$ Cr O$_4$).

The Standardizing.—Now test the first of these three portions of salt with the silver solution, as follows:

Carefully wash a burette; then empty it; then rinse it with a few cubic centimetres of the standard solution. Draw off the rinsings into a waste beaker. Fill the burette to the zero-mark with the standard solution. Next, drop the solution slowly from the burette into the first solution of salt.

At first, a red precipitate of Silver chromate (Ag$_2$ Cr O$_4$) appears.

$$2 \text{ Ag NO}_3 + \text{K}_2 \text{ Cr O}_4 = \text{Ag}_2 \text{ Cr O}_4 + 2 \text{ KNO}_3.$$

But the color is quickly whitened by the formation of Silver chloride (Ag Cl).

$$\text{Ag}_2 \text{ Cr O}_4 + 2 \text{ Na Cl} = 2 \text{ Ag Cl} + \text{Na}_2 \text{ Cr O}_4.$$

Add more Silver nitrate; upon stirring, the red precipitate whitens, but with increasing difficulty, because the amount of common salt (Na Cl) is rapidly diminishing; at length, the common salt being entirely decomposed, the next additional drop of Silver nitrate forms red Silver chromate (Ag$_2$ Cr O$_4$), which is *permanent,* giving a creamy or reddish shade to the whole of the solution. That this is the end of the operation is usually evident from the additional fact that the Silver chloride now collects at the bottom of the solution.

Next, test the second and the third portions of salt, as just described for the first.

The Calculations.—Make an average of the results of the three tests. From the number of cubic centimetres of silver solution used, calculate the value of one cubic centimetre of it, in chlorine, in common salt, and in Potassium chloride. Label the bottle, containing the solution, somewhat as follows, and fill the blank spaces:

STANDARD SILVER SOLUTION FOR **CHLORINE.**

I c. c. ✿ .... m.g. Cl
I c. c. ✿ .... m.g. Na Cl
I c. c. ✿ .... m.g. K Cl

(Date) ....

The Process.

The Weighing.—Weigh about one-half gramme of ordinary or commercial common salt.

The Dissolving.—Dissolve the weighed sample in about 250 cubic centimetres of water. Dilute the solution to exactly 500 cubic centimetres. From this solution, take three successive portions, each measuring 100 cubic centimetres, and test them by means of the standard solution, as described on pp. 93, 94.

The Calculation.—From the number of cubic centimetres used, and from the data already recorded on the label, calculate the amount of Chlorine present in the commercial salt.

Notes.

53. The *standard solution* should be preserved with care. If any water gets into it, its standard will be lowered; if it be so exposed that any water evaporates, its standard will be raised. (See p. 22.)

54. The standard solution must not be added in sufficient quantity to impart a *deep* red color to the solution.

So long as there remains in the solution *any chlorine not combined with silver*, the red precipitate cannot be permanent. When, however, the red precipitate is permanent, it shows that there has been added not only enough of the silver to combine with all the chlorine present, but also the slight excess sufficient to form red Silver chromate.

55. All burettes should be rinsed after using, and should be left full of water.

FOURTEENTH EXERCISE—CHROMIUM.

Data.

	Molecular Weight.	Per Cent.
K$_2$	78.	26.558
Cr$_2$	104.	35.410
O$_7$	111.7	38.032
	293.7	100.000
K$_2$	78.	7.827
Cr$_2$	104.	10.437
O$_3$	47.9	4.807
(SO$_3$)$_3$	239.6	24.044
SO$_3$	79.9	8.018
O	16.	1.606
24 H$_2$O	431.1	43.261
	996.5	100.000
Cr$_2$	104.	68.466
O$_3$	47.9	31.534
	151.9	100.000

The Compound Tested.

The substance tested is crystallized Potassium dichromate ($K_2 Cr_2 O_7$).

Outline of the Process.

Precipitate the Chromium as Chromic hydroxide ($Cr_2 O_6 H_6$); change it, by heating, into Chromic oxide ($Cr_2 O_3$) and weigh it in the latter form.

If the Chromium exists in its higher oxidized forms, it must be *reduced*, prior to precipitation. This *reduction* is accomplished by some deoxidizing agent, of which Ethyl alcohol, $(C_2 H_5)$ O H, Sulphuretted hydrogen $(H_2 S)$, and Sulphurous anhydride (SO_2) are examples. ·

The Process.

The Weighing.—Weigh about one gramme of pure crystallized Potassium dichromate $(K_2 Cr_2 O_7)$.

The Dissolving and Reducing.—Dissolve the salt in a small amount of water. Next add a few drops of pure hydrochloric acid (H Cl).

$$K_2 Cr_2 O_7 + 2 H Cl = 2 Cr O_3 + 2 K Cl + H_2 O.$$

The Chromic anhydride $(Cr O_3)$ liberated, dissolves in the water.[56] Next add cautiously[57] some Ethyl alcohol $(C_2 H_5)$ O H. Boil the solution, and then evaporate it to dryness to expel the excess of hydorchloric acid, as well as the excess of alcohol and the aldehyde[58] formed. Take care not to overheat the residue from the evaporation. Dissolve this residue in water — with help of a drop of hydrochloric acid, if necessary — and boil the solution.

FIG 24.—Convenient Water-bath arrangement for evaporating the Chromic solution to dryness.

' **The Precipitation.**—To the boiling liquid, add Ammonium hydroxide (at first, drop by drop) in slight excess.

Boil the solution until, upon allowing the precipitate to subside for a minute or two, the supernatant liquid appears to be colorless.[60]

$$Cr_2 Cl_6 + 6 (NH_4) O H = Cr_2 O_6 H_6 + 6 NH_4 Cl.$$

The Filtering.—Decant the clear liquid through a filter; wash the precipitate several times by decantation. Finally, transfer the precipitate itself to the filter.

The Burning.—Dry the precipitate; then remove it from the filter to a piece of glazed paper. Burn the filter-paper first; then add the precipitate to the contents of the crucible, and ignite the whole.

$$Cr_2 O_6 H_6 \; heated = Cr_2 O_3 + 3 H_2 O.$$

Notes.

56. It might be expected that addition of strong acid to Potassium dichromate would yield free Chromic acid ($H_2 Cr O_4$). But, in fact, Chromic anhydride ($Cr O_3$) appears to be formed.

57. Chromic anhydride is a very powerful oxidizing agent, and it yields great heat when acting upon organic compounds. When alcohol is added in large quantity to it, it sometimes froths up and overflows the vessel.

58. The reactions of the deoxidizing agents are given below:

For *Alcohol:*

$$2 Cr O_3 + 6 H Cl + 3 (C_2 H_5) O H = Cr_2 Cl_6 + 6 H_2 O + 3 C_2 H_4 O.$$
$$\text{\small\textit{Aldehyde.}}$$

For *Sulphuretted hydrogen:*

$$2 Cr O_3 + 6 H Cl + 3 H_2 S = Cr_2 Cl_6 + 6 H_2 O + 3 S.$$

For *Sulphurous anhydride:*

$$2 Cr O_3 + 3 SO_2 = Cr_2 (SO_4)_3.$$

59. Organic substances often prevent the precipitation of hydrates of metals from their solutions. The alcohol and aldehyde must be removed in order to avoid this sort of influence.

60. Ammonium hydroxide dissolves freshly precipitated Chromic hydrate, forming a violet-red liquid; but upon thorough boiling, the colored compound is decomposed, with liberation of the whole of its Chromic hydrate.

61. Chromic oxide ($Cr_2 O_3$), which has been *strongly* ignited, is insoluble in Hydrochloric acid.

FIFTEENTH EXERCISE—COPPER.
(PRECIPITATION BY IRON.)

Data.

	Molecular Weight.	Per Cent.
Cu	63.2	25.392
O	16.	6.428
SO_3	79.9	32.101
$5 H_2O$	89.8	36.079
	248.9	100.000

The Compound Tested.

This is crystallized Cupric sulphate, known in commerce as Blue vitriol ($Cu SO_4 + 5 H_2 O$).

Outline of the Process.

Introduce a spiral of iron wire into the solution; this precipitates the copper in the metallic form, in which form it is weighed.

(*a*) This process is not applicable in presence of other substances precipitable by iron, viz.: silver, mercury, lead, arsenic, antimony, bismuth, gold.

(*b*) This process is not applicable to solutions containing Nitric acid ($H NO_3$), except after that acid has been removed: even minute quantities of Nitric acid prevent the precipitation of the last portions of copper.

The Process.

The Weighing.—Weigh accurately about 4 grammes of the Cupric sulphate.

The Dissolving.—Dissolve the weighed substance in hot water; then add about 25 cubic centimetres of Hydrochloric acid to the solution.

Preparation of the Iron Wire.—Select a piece of iron wire of about nine inches in length, and about one-eighth of an inch in diameter. Round the ends of the wire with a file. Clean the surface of the wire by rubbing it with a rag dipped in hydrochloric acid. Finally, wash it with water. Bend the wire into a spiral, so that when it stands in the copper solution it may have its upper end in the upper strata of the liquid.

The Precipitation.—Place the spiral in the copper solution,[62] and let the whole stand in a moderately warm place for about three hours,—or until the copper is completely precipitated, and no longer. This end is known to be reached when a clean piece of iron held in the solution for a few moments acquires no red deposit of copper.

$$Cu\ SO_4 + Fe_2 + 2\ H\ Cl = Fe\ SO_4 + Fe\ Cl_2 + Cu + H_2.$$

The Washing.—When precipitation is complete, brush with a feather the adhering copper from the iron wire into the solution; withdraw the wire, washing it with water. Allow the copper to subside for a few minutes; then decant the clear liquid from the first beaker, designated as A, to a second beaker, designated as B. Fill A full of clean water. Allow the precipitate to subside for a few minutes, or until the liquid is clear.

Pour the clear liquid in B away, saving any copper that may be deposited in it. Pour the clear liquid from A into B. Repeat these washing processes several times.

After the wash waters in A and B have been poured off for the last time, pour a few cubic centimetres of alcohol ($C_2\ H_5\ O\ H$) into B, and by its help wash any copper

9 *

contained in B back into A.   When the copper has set-
tled in A, pour the clear alcohol away.[65]

The Drying. — Next dry the copper in A over a warm
(not hot) sand-bath.   When the copper is perfectly dry,
brush it into a watch-glass and quickly weigh it as metallic copper.

Notes.

62. From a little consideration it will appear that in
making this assay it is necessary—after the copper has
once commenced to precipitate—to continue the work
without delay to its conclusion.

Thus it must be remembered that the freshly precipitated metallic copper is likely to become injured upon
exposure even to pure air, much more upon exposure
to an acid atmosphere.   The air turns the copper either
to red Cuprous oxide ($Cu_2 O$), or to black Cupric oxide
($Cu O$).

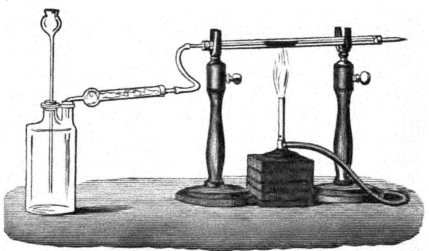

Fig. 25.—Disposition of Apparatus for Reducing Oxidized Copper

Copper which has thus become partly or wholly oxidized may be brought back.   To accomplish this, place
the substance in a small porcelain boat, and the latter in

a combustion-tube of Bohemian glass. Pass a current of hydrogen gas over the copper, and *when the explosive mixture of air and hydrogen* is judged to be expelled, light the hydrogen at the jet. Now heat the tube, where the boat is, to low redness. Presently the oxide will glow, from the heat of the process of reduction. When the reduction is judged to be complete, remove the lamp-flame, but let the hydrogen flow until the copper is cool; now extinguish the hydrogen flame, and take out the boat and weigh it.

63. During the final drying of the copper, the alcohol vapors exert their protective influence in two ways:

First. They fill the beaker and so partly expel the air;

Secondly. Owing to the affinity of alcohol for oxygen, the vapors tend to withdraw oxygen from any air that may be present in the beaker.

SIXTEENTH EXERCISE—COPPER.
(BY ELECTROLYSIS.)

Data.

	Molecular Weight.	Per Cent.
Cu	63.2	25.392
O	16.	6.428
SO_3	79.9	32.101
5 H_2O	89.8	36.079
	248.9	100.000

The Compound Tested.

This is crystallized Cupric sulphate ($CuSO_4 + 5 H_2O$).

Outline of the Process.

Prepare one or two cups of Grove's battery;[64] by their aid, deposit the Copper in the metallic form in a platinum dish. The dish serves not only as a receptacle for the solution to be tested, it is also the negative electrode of the battery; hence it is connected (by a copper wire) with the zinc end of the battery.

The process is applicable to solutions of Copper in absence of Nitric acid, also in absence of substances precipitable by the galvanic current or by metallic Copper.

It should be the invariable rule to have the battery in the best of running order before beginning to plate. With this in mind, observe carefully the directions given.

The Battery.

(*a*) **Provide the Battery Acids.** — Two acids are used. For the inner cell, ordinary concentrated Nitric acid is used. For the outer cup, dilute Sulphuric acid is used. Prepare the dilute Sulphuric acid as follows:

Measure separately in a graduated glass

> 100 parts of water,
>
> 5 parts of concentrated Sulphuric acid.

Pour the water into a beaker; then *slowly pour the acid into the water* and stir the mixture. (The proportions given, represent about one part of acid to ten of water by weight. If this acid affords a precipitate of Lead sulphate, separate the latter from the diluted acid by decantation, or by filtration, before use. Lead sulphate, if present, exercises an injurious local galvanic action upon the zinc.)

(*b*) **Amalgamate the Zincs.**[66]— For this purpose prepare five beakers, each capable of holding one zinc. Fill

— the first, with hot water;

— the second, with dilute Sulphuric acid;

— the third, with water, either hot or cold;

— the fourth, with a solution of Mercuric chloride, acidified with Nitric acid;

— the fifth, with water.

Pass the zincs slowly from one beaker to another, in the order of the numbers, until the surfaces are thoroughly amalgamated.

(*c*) **Soak the Porous Cups.** — This is accomplished by merely filling them with water and allowing them to stand in that condition for a few minutes.

(*d*) **File the Connections.** — This process insures a bright metallic contact where it is necessary, but it must be used with the full appreciation of the fact that unless cautiously performed it rapidly wears away the parts filed.

(*e*) **Set up the Battery.** — One Grove cup[64] is usually sufficient; if two cups are used, they should be arranged for intensity—that is, each zinc should be connected with its neighboring platinum and not with its neighboring zinc.

The Process.

The Weighing and Dissolving. — Prepare and weigh the platinum dish that is to receive the deposited metal. Weigh into this dish about one gramme of crystallized

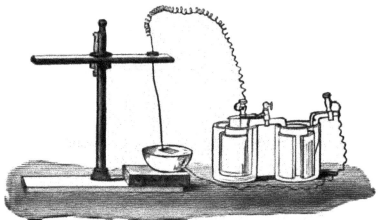

Fig. 26.—Disposition of Apparatus for the Electrolytic Deposition of Copper.

Cupric sulphate. Dissolve the salt in the dish in a suitable quantity of water, and add a drop of pure Sulphuric acid.[68]

The Plating. — Place the platinum dish upon the end of the wire attached to the zinc end of the battery; the wire should be wound into a flat coil for this purpose. Above suspend a platinum electrode[67],—attached to the other end of the battery,—so that it will dip into the copper solution. Deposition of the metal should commence at once; one or two hours should suffice for its completion.

The Washing.—When the process is terminated, pour away the liquid and wash the copper thoroughly, first with cold water and finally with hot. Dry the metal in a water-bath, and then weigh it.[69]

Care of the Apparatus.

Clean the Battery.—Pour away the acids used, and after washing the various parts of the apparatus, set them in place and fill them with water; leave the battery in this condition.

Clean the Platinum Dish.—Warm some Nitric acid in it, and afterward thoroughly wash it with water.

Notes.

64. A single cup of Grove's battery, as ready for action, may be described as follows. The outer vessel is a jar containing dilute Sulphuric acid. In the acid is a hollow cylinder of zinc and a porous earthen cup; the cup contains concentrated Nitric acid, in which a strip of platinum is suspended. A metallic conducting wire attached to the platinum is called the positive electrode of the battery; a similar wire attached to the zinc is called the negative electrode.

Fɪɢ. 27.—Perspective View of a Single Cup of Grove's Battery.

65. While there are several ways of producing a galvanic current, the principal one is by means of a properly regulated chemical action. The conditions under which the current is thus generated may be briefly stated as follows:

First, three substances usually take part; *second,* these

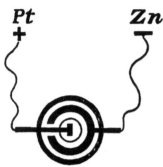

substances must all be conductors of the current; *third,* the chemical action between the substance *a* and the substance *b* must be greater than that between the substance *b* and the substance *c; fourth,* the substances are usually two metals and one liquid, though we are by no means confined to this combination.

Fig. 28.—Diagram Representing the Parts of a Single Cup of Grove's Battery.

As examples of the combinations actually employed, note the following table:

Name.	The Metal consumed.	The Exciting Liquid.	The Third Substance.
Wollaston's	Zinc.....	Sulphuric acid............	Copper.
Smee's.......	Zinc.....	Sulphuric acid............	Platinum.
Walker's.....	Zinc.....	Sulphuric acid............	Carbon.
Grove's......	Zinc.....	Sulphuric acid............	Platinum in Nitric acid.
Bunsen's ...	Zinc.....	Sulphuric acid............	Carbon in Nitric acid.
Leclanche's	Zinc.....	Ammonium chloride....	Carbon with or without Manganese dioxide.
Daniell's ...	Zinc.....	Solution of Cupric sulphate.	Copper.

This brief table at once suggests several facts: *First,* that zinc is very frequently used as the metal to be dissolved by acid in the action of the battery; *second,* that carbon and platinum, being very inert substances, are preferred as the inner elements of the cell; *third,* that the exciting liquids, while capable of much variation, are such as have ready action upon zinc, forming soluble salts with it.

The purpose of the Nitric acid in the inner cell is to oxidize that hydrogen gas that the action of the battery liberates upon the inner element. If this hydrogen is allowed to remain where it collects, it polarizes the platinum or the carbon, and materially diminishes the effectiveness of the battery.

FIG. 29.—Perspective View of a Single Cup of Smee's Battery.

66. The amalgamation of the zincs has a tendency to make the battery more constant. It does this by reason of the facts that *pure* zinc is not subject to the injurious local action that is set up at all points where impurities are deposited, and that the mercury dissolves the pure zinc and brings it to the surface of the plates, leaving the original impurities of the metal in the interior.

67. The platinum electrode is easily made as follows:

FIG. 30.—Perspective View of Two Cups of Bunsen's Battery.

Have a gas flame so placed that it streams over a small anvil; next, hold a piece of platinum foil and a piece of platinum wire in this flame, so that both may be heated red-hot at the same time. When both are ready, give them a sudden blow with a hammer; this should strike them down upon the anvil and easily weld them together.

68. Sulphuric acid increases the electrical conductivity of the solution, and so favors the precipitation of the copper.

69. Care must be taken in drying the copper. It must not be dried in an acid atmosphere, for then the metal will be attacked, and so will give rise to too high results. The deposit must not be heated too strongly, for then there is danger not only of changing the metal into either black oxide ($Cu\,O$) or into red oxide ($Cu_2\,O$); there is the additional danger of ruining the platinum dish. (See page 74.)

SEVENTEENTH EXERCISE—COPPER.

(AS BLACK OXIDE.)

Data.

	Molecular Weight.	Per Cent.
Cu	63.2	25.392
O	16.	6.428
SO_3	79.9	32.101
$5 H_2O$	89.8	36.079
	248.9	100.000
Cu	63.2	79.798
O	16.	20.202
	79.2	100.000

The Compound Tested.

This is crystallized Cupric sulphate ($Cu SO_4 + 5 H_2 O$).

Outline of the Process.

Precipitate the copper as Cupric hydroxide; then by ignition change the hydroxide into Cupric oxide ($Cu O$), which is the substance to be weighed.

The Process.

The Weighing.—Weigh about one gramme of the crystallized and pure salt.

The Dissolving. — Dissolve the weighed substance in hot water and bring the solution to the boiling-point.

The Precipitation. — Continue heating the liquid, and then cautiously add solution of Sodium hydroxide until the whole is just alkaline to litmus-paper.[70] The precipitate, at first blue, should become almost black, and should quickly subside when the lamp is withdrawn.

$$Cu\ SO_4 + 2\ Na\ OH = Cu\ O_2 H_2 + Na_2\ SO_4.$$
$$3\ Cu\ O_2\ H_2\ \textit{boiled} = Cu_3\ O_4\ H_2 + 2\ H_2\ O.$$

The Washing. — Pour the clear liquid through the filter. Wash the precipitate repeatedly with hot water, by decantation, for the purpose of removing all the Sodium compounds, some of which adhere to the precipitate with great tenacity.

The Burning. — After thorough drying, separate the precipitate from the paper, and burn the paper first, in a porcelain crucible. It is always desirable to add to the filter-ash a drop of Nitric acid, and then to evaporate to dryness and ignite, before adding the mass of the precipitate.[72] Finally, introduce the entire precipitate and strongly ignite it;[73] the residue should consist solely of Cupric oxide plus filter-ash.

$$Cu_3\ O_4\ H_2\ \textit{heated} = 3\ Cu\ O + H_2\ O.$$

Notes.

70. When Sodium hydroxide is added to a cold solution of Cupric sulphate it at first affords a blue precipitate of Cupric hydroxide ($Cu\ O_2\ H_2$). This precipitate may be viewed as $Cu\ O + H_2\ O$; it is flocculent and difficult to wash; upon boiling, however, it parts with some of its water and becomes a basic hydroxide ($Cu_3\ O_4\ H_2$), which may be viewed as $3\ Cu\ O + H_2\ O$, and which is granular and more easily washed. As analogous to the facts here presented, it should be remembered that a crystalline sub-stance separating from its water solution at a high temperature usually has less water of crystallization than the same salt when crystallized at a low temperature ; further rise of temperature tends to lessen the hydration of soluble salts even while they are still in solution.

71. In presence of non-volatile organic matters all the copper in a solution cannot be precipitated by Sodium hydroxide. (For analogous fact, see Note 59.)

72. Upon burning the filter-paper, its carbon and hydrogen are certain to reduce to the metallic form any Cupric oxide adhering to it. Nitric acid changes the copper to a nitrate, which by ignition is turned back to oxide.

73. Cupric oxide absorbs moisture from the atmosphere, but it absorbs less if it has been strongly ignited. This fact must be borne in mind in weighing the substance.

10*   H

EIGHTEENTH EXERCISE—IRON.
(GRAVIMETRICALLY.)

Data.

	Molecular Weight.	Per Cent.
Fe	55.9	14.278
O	16.	4.087
SO_3	79.9	20.409
$(NH_4)_2$	36.	9.195
O	16.	4.087
SO_3	79.9	20.409
$6 H_2O$	107.8	27.535
	391.5	100.000
Fe_2	111.8	70.006
O_3	47.9	29.994
	159.7	100.000

The Compound Tested.

This is Ammonio-ferrous sulphate $(NH_4)_2 SO_4 + Fe SO_4 + 6 H_2O$. This compound must be carefully distinguished from the Ammonio-ferric sulphate. The latter compound is Iron alum; the former compound is not. (See pages 47, 48.)

Outline of the Process.

Oxidize the iron by boiling with *aqua-regia*. Precipitate the iron as Ferric hydroxide; weigh it as Ferric oxide.

The Process.

The Weighing.—Weigh about one gramme of Ammonio-ferrous sulphate, $(NH_4)_2 SO_4 + Fe SO_4 + 6 H_2 O$.

The Dissolving.—Dissolve the salt in water; add a few drops of pure hydrochloric acid and a few drops of pure Nitric acid; boil to change the iron into the ferric form.

$$6 [(NH_4)_2 SO_4 + Fe SO_4] + 6 H Cl + 6 H NO_3 =$$
$$6 (NH_4)_2 SO_4 + 2 Fe_2 (SO_4)_3 + Fe_2 Cl_6 + 3 N_2 O_4 + 6 H_2 O.$$

The Precipitation.—Now carefully add sufficient Ammonium hydroxide to produce alkaline reaction. The precipitate, which should be dark red, is Ferric hydroxide ($Fe_2 O_6 H_6$),

$$Fe_2 (SO_4)_3 + Fe_2 Cl_6 + 12 (NH_4) OH =$$
$$2 Fe_2 O_6 H_6 + 3 (NH_4)_2 SO_4 + 6 NH_4 Cl.$$

Boil thoroughly; allow the precipitate to subside.[75]

The Filtration.—Pour the clear liquid into the filter; continue the washing by decantation. Finally transfer the precipitate itself to the filter.

The Burning.—Dry the precipitate *thoroughly.* (See page 50, note 5.) Ignite separately the filter and the precipitate (the filter first). The residue is Ferric oxide ($Fe_2 O_3$),

$$Fe_2 O_6 H_6 \text{ } heated = Fe_2 O_3 + 3H_2 O.$$

Notes.

74. The following is a type reaction for *aqua-regia:*

$$2 H NO_3 + 2 H Cl = Cl_2 + N_2 O_4 + 2 H_2 O.$$

Under varying conditions the reaction varies; below is another form of it* involving the production of Nitrosyl chloride (NOCl):

$$H NO_3 + 3 H Cl = 2 H_2 O + NOCl + Cl_2.$$

75. When the iron of the original salt to be tested is in the ferric form, addition of *aqua-regia* is unnecessary.

When the metal is in the ferrous form, addition of

---

* Roscoe & Schorlemmer's *Chemistry*, Vol. I., pp. 412 and 425.

Ammonium hydroxide produces a bluish precipitate of Ferrous hydroxide (Fe O_2 H_2). This compound, upon exposure to air, changes in part to Ferric hydroxide, and thus gives a precipitate of variable constitution.

When the metal is partly ferrous and partly ferric, addition of Ammonium hydroxide produces a black precipitate of variable composition, but containing Ferroso-ferric oxide, also called magnetic oxide (Fe_3 O_4).

76. Ferric hydroxide (Fe_2 O_6 H_6), though soluble in acids, is insoluble in Ammonium hydroxide and Ammonium salts. When precipitated by Sodium hydroxide (Na OH), or by Potassium hydroxide (K O H), the Ferric hydroxide holds part of the alkali with such tenacity that washing will not remove it. Ammonium hydroxide is also retained in a similar manner, but being volatile, it is less objectionable on this account.

77. Ferric oxide (Fe_2 O_3) remains unchanged by ignition in the air ; but at very high temperatures and away from access of air, it changes to Ferroso-ferric oxide (Fe_3 O_4) with loss of oxygen.

78. When ignited in presence of Ammonium chloride (NH_4 Cl), Ferric oxide suffers loss, owing to volatilization of Ferric chloride (Fe_2 Cl_6).

NINETEENTH EXERCISE—IRON.

(THE POTASSIUM PERMANGANATE TEST.)

Data.

	Molecular Weight.	Per Cent.
Fe_2	111.8	11.617
O_3	47.9	4.977
$3\ SO_3$	239.7	24.906
$(NH_4)_2$	36.	3.741
O	16.	1.663
SO_3	79.9	8.302
$24\ H_2O$	431.1	44.794
	962.4	100.000

The Compound Tested.

This is Ammonio-ferric alum, the double Sulphate of iron and ammonium, $(NH_4)_2\ SO_4 + Fe_2\ (SO_4)_3 + 24\ H_2\ O$. (See page 48.)

The process is applicable to almost all compounds of Iron; but (*a*) compounds in the ferric form must be *reduced*, as a part of the process, to the ferrous form; and (*b*) if a solution contains any substance that has a reducing action upon Potassium permanganate, such substance must be removed before this test can be applied.

Outline of the Process.

(The test involves the facts that Potassium permanganate $(K_2\ Mn_2\ O_8)$ is a powerful oxidizing agent and that it has a high color. Now when the permanganate comes in contact with a Ferrous salt, it parts with some of its oxygen to produce a Ferric salt; by reason of this loss of oxygen a colorless compound of manganese results.)

Reduce the iron, to be tested, to the ferrous form, provided it is not already in that form.

Prepare a standard solution of Potassium permanganate.

To the Ferrous solution, to be tested, add from a burette the required quantity of the standard solution of Potassium permanganate. As the strength of the permanganate solution is known, the number of cubic centimetres used gives at once a datum for the determination of the amount of iron present.

The Standard Solution.

The Weighing. — Prepare the solution of Potassium permanganate as follows: Dissolve 1.5 grammes of crystallized Potassium permanganate in a small amount of distilled water, and then dilute the solution to the volume of 500 cubic centimetres. This solution should have a strength such that one cubic centimetre is sufficient to oxidize about 5 milligrammes of metallic Iron. But the *exact* strength of the solution must be determined by standardizing it by means of a weighed amount of pure Iron in the ferrous form.

The Standardizing. — Weigh accurately 2.5 grammes of Ammonio-ferrous sulphate; dissolve it in a small amount of water and dilute the solution to the volume of 500 cubic centimetres.

Take 100 cubic centimetres of this Ferrous solution, transfer it to a beaker, and to it add about 10 cubic centimetres of dilute Sulphuric acid. From a burette carefully drop the Permanganate solution, little by little, into the Ferrous solution until the latter retains a slightly pinkish .tint. The pink color indicates that a slight excess of permanganate has been added, and that, of the total quantity employed, the principal portion has been decolorized as a result of its oxidizing work on the Ferrous salt. Thus the fact that this required work has been fully accomplished is proved by the colored, and hence undecomposed, condition of the slight excess of permanganate still seen in the solution.

Repeat this operation four times, using 100 cubic centimetres of the Ferrous solution each time.

$$K_2 Mn_2 O_8 + 3 (H_2 SO_4) = K_2 SO_4 + 2 Mn SO_4 + 3 H_2 O + 5 O.$$
$$10 Fe SO_4 + 5 O + 5 H_2 SO_4 = 5 [Fe_2 (SO_4)_3] + 5 H_2 O.$$

The Calculation.—Note the number of cubic centimetres of Permanganate so used. From an average of the *good* results estimate the value in Iron of each cubic centimetre of the Permanganate solution.

The Process.

The Weighing. — Weigh 5 grammes of Ammonioferric alum for testing.

The Dissolving. — Dissolve the weighed salt in water —adding some sulphuric acid if necessary—and make the solution up to the volume of 500 cubic centimetres.

The Reducing.—Into three separate beakers, measure portions of the solution, 100 cubic centimetres to each. Next reduce the iron in each solution to the Ferrous form. To accomplish this reduction add to each portion of the solution about 50 cubic centimetres of dilute Sulphuric acid and then a fragment of platinum foil or a platinum dish, and *in contact* with the foil a piece of pure but amalgamated zinc.[79] Rapid evolution of hydrogen ensues, and the liquid soon becomes colorless.

$$Zn + H_2 SO_4 = Zn SO_4 + H_2.$$
$$Fe_2 (SO_4)_3 + H_2 = 2 Fe SO_4 + H_2 SO_4$$

The end of the operation must be judged by testing a single drop of the solution, upon a piece of white porcelain, with a drop of Potassium ferrocyanide($K_4 Fe Cy_6$). When all the iron is changed to the ferrous form, this ·test fails to yield Prussian blue ($Fe_4 Fe_8 Cy_{18}$).* It is also advisable to test the solution with Potassium sulphocyanate (K S Cy).

When the reduction is judged to be complete, carefully withdraw the zinc and platinum.

The Titration. — Proceed to test separately each of the three solutions that have been reduced, making the tests by dropping in permanganate solution until the pink color appears.

The Calculation. — Multiply the number of cubic centimetres of permanganate solution

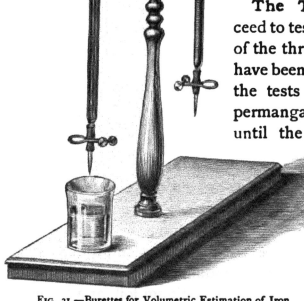

Fig. 31.—Burettes for Volumetric Estimation of Iron.

used, into the value of one cubic centimetre of it expressed

---

* Appleton's *Qualitative Analysis*, pp. 36 and 38.

in iron, and thus estimate the amount of iron in the substance tested.

Notes.

79. Platinum and amalgamated zinc give rise to a galvanic action (see page 107); by reason of this action hydrogen is given off with great freedom. The zinc in excess should hold together so that it may be finally withdrawn as a whole, and thus save the annoyance caused by fragments of zinc; if such fragments appear in the solution, they must be dissolved. In dissolving, there is danger either of oxidizing the reduced iron or else of leaving some fragments of zinc; these latter reduce the permanganate, and so give results that are too high.

80. The use of hydrochloric acid in large quantity is not admissible in this process; in concentrated solutions hydrochloric acid reduces the Permanganic acid, and thus erroneously gets counted as Iron.

$$K_2Mn_2O_8 + 16\,HCl = 2\,KCl + 2\,MnCl_2 + 8\,H_2O + 10\,Cl.$$

81. By the action of Permanganate on Ferrous solutions a brown precipitate of Manganese dioxide is sometimes formed; an excess of Sulphuric acid dissolves this precipitate.

$$K_2Mn_2O_8 + 4\,H_2SO_4 + 6\,FeSO_4 =$$
$$2\,MnO_2 + 3\,Fe_2(SO_4)_3 + K_2SO_4 + 4\,H_2O.$$
$$MnO_2 + H_2SO_4 = MnSO_4 + H_2O + O.$$

82. The permanganate must be added at a tolerably rapid rate. At the end of the process the pink color formed disappears of itself; hence the last stages of the operation must not proceed too slowly, for thus the permanganate might be added indefinitely without producing a permanent pink color.

TWENTIETH EXERCISE—IRON.
(THE STANNOUS CHLORIDE TEST.)

Data.

	Molecular Weight.	Per Cent.
Fe	55.9	14.278
O	16.	4.087
SO_3	79.9	20.409
$(NH_4)_2$	36.	9.195
O	16.	4.087
SO_3	79.9	20.409
$6 H_2O$	107.8	27.535
	391.5	100.000

The Compound Tested.

This is Ammonio-ferrous sulphate $(NH_4)_2 SO_4 + Fe SO_4 + 6 H_2O$. This compound must be carefully distinguished from the Ammonio-ferric sulphate, Iron alum. (See pages 47, 48.)

Outline of the Process.

(*a*) Prepare the standard solutions required. Make a tabular statement, showing how many milligrammes of metallic Iron is represented by one cubic centimetre of each of the standard solutions used.

(*b*) Oxidize the Iron to be tested to the ferric form; then heat the solution to boiling.

(*c*) Into the hot solution, which has a deep yellow color, draw a *slight* excess of standard Stannous chloride solution from a burette.

(*d*) Cool the colorless solution produced by (*c*), and add a few cubic centimetres of Starch liquor.

(*e*) By means of the solution of iodine estimate the excess of Tin solution added.

(*f*) Make the necessary calculations.

The Standard Solutions.

Prepare solutions as follows:

1. **The Ferric Solution.**—This is a standard solution of Ammonio-ferric sulphate, Iron alum, $(NH_4)_2 SO_4 + Fe_2 (SO_4)_3 + 24 H_2 O$. Dissolve 10.757 grammes of the crystallized salt[83] in a small amount of hot water, adding a little hydrochloric acid if necessary; dilute the solution to the volume of 250 cubic centimetres. One cubic centimetre of this solution contains the equivalent of 5 milligrammes of metallic iron.[83]

2. **The Stannous Solution.**—This is a solution of Stannous chloride,[89] Tin crystals $(Sn Cl_2 + 2 H_2 O)$. Dissolve about 5 grammes of the crystals in pure hydrochloric acid;[87] dilute the solution to the volume of 1 litre.

3. **The Iodine Solution.**—Weigh[84] about 500 milligrammes of the purified Iodine;[85] dissolve it in water by aid of a few crystals of Potassium iodide; dilute the solution to the volume of 250 cubic centimetres.

4. **A Thin Paste of Starch.**—Prepare this by softening about 3 grammes of Starch in 100 cubic centimetres of boiling water; allow the mixture to cool.

The Value of the Iodine Solution.

Determine the relation between the Tin solution and the Iodine solution. Proceed as follows: Draw from a burette, into a clean casserole or beaker, 10 cubic centimetres of the Tin solution; add 5 cubic centimetres of

the Starch paste; into the mixture draw Iodine solution
from another burette until the blue color produced
remains permanent after stirring. From the number of
cubic centimetres of Iodine solution used the value of
one cubic centimetre of it, in terms of the Tin solution,
may be calculated.

$$Sn\ Cl_2 + I_2 + 2\ H\ Cl = Sn\ Cl_4 + 2\ HI.$$

The Value of the Stannous Solution.

Take 20 cubic centimetres of the standard Ferric solution (prepared as above, page 123); place them in a casserole with 20 cubic centimetres of pure concentrated hydrochloric acid, and boil; while the solution is boiling,[83]
run in, from a burette, the Stannous chloride solution until
the yellow color of the Iron solution is wholly destroyed.

$$Fe_2(SO_4)_3 + Sn\ Cl_2 + 2\ H\ Cl = 2\ Fe\ SO_4 + Sn\ Cl_4 + H_2\ SO_4.$$

Place the casserole, with its contents, in a basin of cold
water to cool. When cold,[86] add 5 cubic centimetres of
the Starch paste, and run in the Iodine solution until the
blue color is produced.

Knowing the value of the Iodine solution in terms of
the Tin solution, the exact number of cubic centimetres
of the Tin solution required for 20 cubic centimetres of
the Ferric solution may be calculated.

Repeat the operation with three different portions, of
20 cubic centimetres each, of the Ferric solution.

From the average number of cubic centimetres of the
Tin solution used find the value of one cubic centimetre
of the Tin solution in terms of Iron solution, and from
that, by calculation, its value in terms of metallic Iron.
Work out the necessary figures for filling the blanks in
the following table:

Results of Standardizing.

10 c. c. Sn Cl$_2$ solution $= \ldots$ c. c. Iodine solution.
.. c. c. Sn Cl$_2$ solution $= \quad$ 1 c. c. Iodine solution.

20 c. c. Iron solution $= \ldots$ c. c. Sn Cl$_2$ solution.
       Subtract $\qquad \ldots$ c. c. Sn Cl$_2$ solution, $\left.\begin{array}{l}\\ \text{the equivalent} \\ \text{of} \ldots \ldots \text{c. c.} \\ \text{Iodine solution} \\ \text{run back.}\end{array}\right\}$

20 c. c. Iron solution $=$ *net* $\ldots$ c. c. Sn Cl$_2$ solution.

SUMMARY.

but   20 c. c. Iron solution $=$ 100 milligrammes of Iron.
hence 1 c. c. Sn Cl$_2$ solution $= \ldots$ milligrammes of Iron.
      1 c. c. Iodine solution $= \ldots$ milligrammes of Iron.

The Process.

The Weighing. — Weigh 2.500 grammes of Ammonio-ferrous sulphate, $(NH_4)_2 SO_4 + Fe SO_4 + 6 H_2 O$.

The Dissolving. — Dissolve the salt in water; add pure concentrated hydrochloric acid and a **very few** crystals of Potassium chlorate ($KClO_3$). For this purpose weigh about 100 milligrammes of the chlorate and add it—crystal by crystal as needed—to the Iron solution until the latter is oxidized.[90]

Boil the solution for some time, in order both to help oxidize the iron from the ferrous to the ferric form and to expel oxides of chlorine.

$$3 Fe SO_4 + Fe Cl_2 + 3 K Cl O_3 + 10 H Cl =$$
$$Fe_2 (SO_4)_3 + Fe_2 Cl_6 + Cl_2 O_4 + 3 K Cl + 5 H_2 O + Cl_4.$$

Dilute the solution to the volume of 500 cubic centimetres.

11 *

The Titration. — Place in a casserole 100 cubic centimetres of this ferric solution; add 20 cubic centimetres of pure concentrated hydrochloric acid, and boil the whole. While boiling, run in the Tin solution from a burette until the Iron solution becomes colorless. Now cool the colorless solution; when it is cold, estimate the *excess* of Tin solution as follows:

Add to the colorless solution 5 cubic centimetres of Starch paste; then draw from a burette the Iodine solution, drop by drop, until the blue color of Iodine and Starch appears.

Repeat the operations with three different portions of the Iron solution to be tested.

The Calculation. — Find, as above directed, the average number of cubic centimetres of the Tin solution actually needed for the Iron. Knowing the value of 1 cubic centimetre of the Tin solution in terms of metallic Iron (see page 125), the number of milligrammes of metallic Iron in 100 cubic centimetres of the Iron solution tested may be obtained by a simple multiplication; by an easy calculation the *percentage amount* may be found.

Notes.

83. The strength of the standard Iron alum solution should be such that one cubic centimetre contains 5 milligrammes of metallic iron. The amount of the crystallized salt needed to furnish this amount of iron in a solution of 250 cubic centimetres is found by the following proportion:

Molecular weight of Iron		Molecular weight of Iron alum		Grammes of Iron		Grammes of Iron alum
111.8	:	962.4	: :	1.250	:	10.761

84. Iodine is constantly evolving vapors which corrode the metal work of the balances. The weighing should therefore be quickly performed and the balance-case aired afterwards.

85. Pure Iodine is not absolutely necessary in this analysis. Iodine may be purified, when desired, as follows:

Take about 3 grammes of Iodine, rub it in a mortar with a few crystals of Potassium iodide (KI). Place the mixture between two watch-glasses, and gently heat so as to vaporize the Iodine. Allow the glasses to cool, and when cold, scrape off the resublimed Iodine from the upper glass. The Potassium iodide purifies the Iodine from chlorine and bromine.

$$KI + Cl = K\ Cl + I.$$
$$KI + Br = K\ Br + I.$$

86. Heat destroys the blue color of the Iodide of starch, hence the solution must be cooled before adding the iodine.

87. When Tin crystals are added to water, there sometimes appears a basic salt of Tin, which is almost completely insoluble in water and difficultly soluble even in dilute hydrochloric acid. The formation of this salt is prevented by dissolving the Tin crystals at once in boiling pure concentrated hydrochloric acid and afterward diluting the solution with warm water.

88. The Iron solution must be at or near the boiling point while adding the Tin solution, otherwise the reaction is not energetic and complete.

89. It must be remembered that the Tin solution is not permanent, consequently it must be tested afresh from time to time.

90. To ascertain that all the iron is oxidized to the ferric form, it is sufficient to show that ferrous iron is no longer present. For this purpose test the liquid by placing a drop of it in a drop of solution of Potassium ferricyanide; no blue color should appear, for the reagent gives only a brown color with ferric salts, while it gives a deep blue precipitate (a variety of Prussian blue) with ferrous salts. But the solution of Potassium ferricyanide must be freshly prepared, as it decomposes upon keeping.

TWENTY-FIRST EXERCISE—LEAD.

Data.

	Molecular Weight.	Per Cent.
Pb	206.5	62.519
O	16.	4.844
N_2	28.	8.477
O_5	79.8	24.160
	330.3	100.000
Pb	206.5	68.287
O	16.	5.291
SO_3	79.9	26.422
	302.4	100.000

The Compound Tested.

This is crystallized Lead nitrate, $Pb(NO_3)_2$.

Outline of the Process.

Precipitate the lead by addition of dilute Sulphuric acid, in presence of Ethyl alcohol (C_2H_5OH), as Lead sulphate ($PbSO_4$). Weigh the latter compound on a balanced filter.

The Process.

The Weighing. — Weigh one gramme of Lead nitrate, $Pb(NO_3)_2$.

The Dissolving. — Dissolve the weighed salt in a small amount of distilled water.

The Precipitation. — To the solution add a slight excess of dilute Sulphuric acid. A white precipitate of Lead sulphate ($PbSO_4$) appears.

$$Pb(NO_3)_2 + H_2SO_4 = PbSO_4 + 2HNO_3.$$

I

Now add an amount of Ethyl alcohol, ($C_2 H_5 OH$), equal in bulk to the present volume of the solution.[91]

The Filtration. — When the precipitate has completely subsided, decant the clear liquid upon a balanced filter.[9] Wash the precipitate thoroughly with Spirit of wine,[92] and finally transfer it to the filter. Save the filtrate, and if clear, pass it through the counterbalancing filter, finally washing the latter with a little more Spirit of wine.

The Weighing. — Dry both filters. Find the weight of the Lead sulphate by placing on one balance-pan the filter containing the precipitate, and upon the other pan the counterbalancing filter, together with a sufficient quantity of weights.

Notes.

91. Lead sulphate is slightly soluble in pure water, but considerably less so in water containing Sulphuric acid; but *concentrated* Sulphuric acid dissolves it.

In alcohol, Lead sulphate is almost completely insoluble. Presence of Nitric or of Hydrochloric acids increases its solubility.

Ammonium salts, especially Nitrate, Acetate, and Tartrate, readily dissolve Lead sulphate.

92. Spirit of wine is 80 per cent. Alcohol. It may be prepared by adding to strong alcohol about one-fourth its volume of water.

93. Thorough washing is necessary, since, if any free Sulphuric acid remains in the filter-papers, it concentrates upon drying and chars the paper.

TWENTY-SECOND EXERCISE—LEAD.
(METHOD FOR GALENA.)

Data.

	Molecular Weight.	Per Cent.
Pb	206.5	86.583
S	32.	13.417
	238.5	100.000
Pb	206.5	68.287
O	16.	5.291
SO$_3$	79.9	26.422
	302.4	100.000

The Compound Tested.

This is Galena, natural Lead sulphide (Pb S). It occurs in Nature in cubically crystalline masses, which are sometimes free from all impurity.

Outline of the Process.

(*a*) By means of fuming Nitric acid (page 56), oxidize the Lead sulphide (Pb S) into Lead sulphate (Pb SO$_4$). Weigh the insoluble oxidized product upon a balanced filter.

(*b*) Digest the weighed product in Ammonium acetate solution ; this dissolves Lead sulphate.

(*c*) Weigh the insoluble matter, and find the amount of Lead sulphate by difference.

The Process.

The Weighing. — Weigh about one gramme of the very finely pulverized ore.

The Oxidizing. — Place the weighed portion in a dry beaker, add a few cubic centimetres of fuming Nitric acid, and cover the beaker with a large watch-glass. After allowing the reaction to go on in the cold for a few minutes, gently warm the solution.

$$3 \, Pb \, S + 16 \, H \, NO_3 =$$
$$Pb \, SO_4 + 2 \, [Pb \, (NO_3)_2] + 2 \, S + 6 \, N_2 \, O_4 + 8 \, H_2O.$$

The Precipitation. — When the reaction seems to be complete, add a little dilute Sulphuric acid;[96] then carefully evaporate the whole to dryness. If the operation is properly performed, no globules of free Sulphur will appear at this stage.

Treat the cooled, evaporated residue with dilute Sulphuric acid and a little water, and then add some Spirit of wine.

The Filtration. — Filter on a balanced filter, dry and weigh as directed on page 130.

The Gangue. — After the weight of the precipitate is taken, carefully remove the latter from its paper and digest it in Ammonium acetate, [96] $(NH_4) \, O \, (C_2 \, H_3 \, O)$. The Lead sulphate will be dissolved by this treatment. Filter; the gangue (Silica, Si O_2, etc.,) will be left upon the filter. Wash with hot water, dry, and ignite.

The Calculation. — Subtract the weight of gangue from the combined weight of Lead sulphate and gangue previously obtained; from the difference estimate the amount of lead.

Notes.

94. Ordinarily, Galena contains some gangue and also a trace of silver; those samples that are distinctly crystallized usually contain less silver.

95. Fuming Nitric acid[11] oxidizes most of the Lead sulphide into Lead sulphate, but a little Lead nitrate Pb $(NO_3)_2$, is formed. The Sulphuric acid is added for the purpose of changing this Nitrate into the Sulphate.

96. To prepare Ammonium acetate, take a suitable quantity of Ammonium hydroxide, and to it add just sufficient Acetic acid to afford acid reaction. Then add more Ammonium hydroxide, sufficient to produce decided alkaline reaction.

12

TWENTY-THIRD EXERCISE—MAGNESIUM.

Data.

	Molecular Weight.	Per Cent.
Mg	24.	9.772
O	16.	6.515
SO_3	79.9	32.533
$7\ H_2O$	125.7	51.181
	245.6	100.001
Mg_2	47.9	21.625
O_2	31.9	14.402
P_2	61.9	27.946
O_5	79.8	36.027
	221.5	100.000

The Compound Tested.

This is crystallized Magnesium sulphate, Epsom salts, $(Mg\ SO_4 + 7\ H_2O)$.

Outline of the Process.

Precipitate the Magnesium as Ammonio-magnesium phosphate $(NH_4)\ Mg\ (PO_4)$. Afterward heat the precipitate to redness. Weigh the substance as Magnesium pyro-phosphate $(Mg_2\ P_2\ O_7)$.

The Process.

The Weighing. — Weigh one gramme of crystallized Magnesium sulphate $(Mg\ SO_4 + 7\ H_2O)$.

The Dissolving. — Dissolve the weighed salt in water and add a few drops of Hydrochloric acid. Add a small quantity of a solution of Ammonium chloride $(NH_4)\ Cl$, and then Ammonium hydroxide $(NH_4\ OH)$, to alkaline reaction.

The Precipitation.—To the Magnesium solution add a slight excess of a solution of Hydro di-sodium phosphate ($H\ Na_2\ PO_4$).

$$Mg\ SO_4 + (NH_4)\ OH + H\ Na_2\ PO_4 =$$
$$(NH_4)\ Mg\ PO_4 + Na_2\ SO_4 + H_2\ O.$$

Stir the solution, taking care not to allow the glass rod to touch the sides of the beaker, otherwise lines of crystals not easily detached are apt to be formed. Allow the precipitate to stand for from twelve to twenty-four hours.

The Filtration.—Transfer the clear liquid to a filter; wash the precipitate by decantation with a solution of three parts water and one part Ammonium hydroxide.[97] Finally place the precipitate on the filter.

The Burning. — After drying ignite the precipitate in a platinum crucible, burning the paper first. After heating for some time over the ordinary lamp, heat the precipitate for a few minutes with the blast-lamp.[98] The precipitate, which consisted of Ammonio-magnesium phosphate, $(NH_4)\ MgPO_4$, is changed by the ignition into Magnesium pyro-phosphate ($Mg_2P_2O_7$), and is weighed as such.

$$2\ [(N\ H_4)\ Mg\ P\ O_4]\ heated = Mg_2\ P_2\ O_7 + 2\ NH_3 + H_2\ O.$$

Notes.

97. The precipitate of Ammonio-magnesium phosphate is somewhat soluble in water, but less so in presence of Ammonium salts. The *crystallized* precipitate has, before drying, the formula,

$$(NH_4)\ Mg\ PO_4 + 6\ H_2\ O.$$

98. If the Pyro-phosphate fuses in the crucible, it may be removed by fusing it afterward with a mixture of Sodium and Potassium carbonates, followed, after cooling, by digestion in water. With respect to the care of platinum vessels, see page 74.

TWENTY-FOURTH EXERCISE—MERCURY.

Data.

	Molecular Weight.	Per Cent.
Hg	199.7	73.854
Cl$_2$	70.7	26.146
	270.4	100.000
Hg	199.7	86.189
S	32.	13.811
	231.7	100.000

The Compound Tested.

This is Mercuric chloride, Corrosive sublimate (Hg Cl$_2$).

Outline of the Process.

Precipitate the Mercury as Mercuric sulphide (Hg S). Dry the sulphide at 212° F., and weigh it.

The Process.

The Weighing.—Weigh one gramme of the Mercuric chloride.

The Dissolving.—Dissolve the weighed salt in water acidulated with hydrochloric acid; then dilute considerably[100] with water.

The Precipitation.—Pass a stream of hydrosulphuric acid gas through the solution until precipitation is complete.

$$\text{Hg Cl}_2 + \text{H}_2\text{ S} = \text{Hg S} + 2 \text{ H Cl.}$$

Allow the solution to stand a few minutes for the precipitate to subside.

The Filtration.—Pass the clear liquid through a balanced filter; then transfer the precipitate to the same; quickly wash with cold water.

The Drying.—Dry the precipitate at 212° F., and weigh it.

If, from any cause, (presence of Ferric oxide, free Chlorine, or the like,) the precipitate should contain free sul-

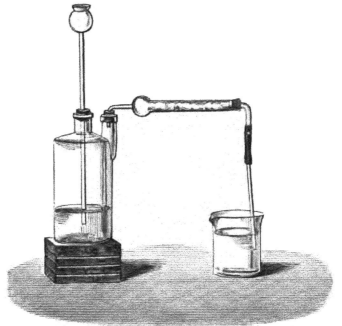

Fig. 32.—Disposition of Apparatus for Precipitation of Mercuric Sulphide.

phur, this sulphur may be removed by passing a little Carbon disulphide through the filter containing the dried precipitate.

Note.

100. From a solution of Mercuric chloride, containing much free hydrochloric acid, the whole of the metal cannot be precipitated as sulphide by means of Sulphuretted hydrogen — until the solution is properly diluted. If the experiment is made with *concentrated* solutions, the precipitate may contain Mercurous chloride ($Hg_2 Cl_2$) and free sulphur, as well as variable mixtures of Mercuric chloride and Mercuric sulphide.

12*

Twenty-Fifth Exercise—Nickel.
(By Electrolysis.)

Data.

	Molecular Weight.	Per Cent.
Ni	57.9	14.714
O	16.	4.066
SO_3	79.9	20.305
$(NH_4)_2$	36.	9.148
O	16.	4.066
SO_3	79.9	20.305
$6 H_2O$	107.8	27.395
	393.5	99.999

The Compound Tested.

The substance tested is the double Sulphate of Nickel and Ammonia, also called Ammonio-nickelous sulphate, $(NH_4)_2 SO_4 + Ni SO_4 + 6 H_2O$.

Outline of the Process.

Prepare two cups of Grove's battery;[64] by their aid deposit the nickel, in the metallic form, in a platinum dish. The dish serves not only as a receptacle for the solution to be tested, it is also the negative electrode of the battery; hence it is connected (by a copper wire) with the zinc end of the battery.

It should be the invariable rule to have the battery in the best of running order before beginning to plate. With this in mind, observe carefully the directions given here and at pages 105 and 106.

The Battery.

(*a*) Provide the battery acids. (See page 105.)

(*b*) Amalgamate the zincs.

(*c*) Soak the porous cups.

(*d*) File the connections.

(*e*) Set up the battery. (See page 106.)

The Process.

The Weighing and Dissolving.—Prepare the platinum dishes that are to receive the deposited metal. Weigh into the weighed platinum dish about one gramme of the double Sulphate of Nickel and Ammonium. Dissolve the salt in the dish in about 15 cubic centimetres of water, and add sufficient Ammonium hydroxide to produce strong alkaline reaction.

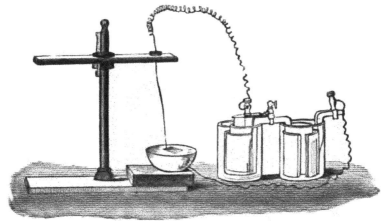

FIG. 33.—Disposition of Apparatus for the Electrolytic Deposition of Nickel.

The Plating.—Place the platinum dish upon the end of the wire attached to the zinc end of the battery; the wire should be wound into a flat coil for this purpose. Above suspend a platinum electrode,[67] attached to the other end of the battery, so that it dips into the nickel solution. Deposition of the metal should commence at once. Continue the plating until all the nickel has been deposited. We may judge this stage to be reached when the solution—being still distinctly alkaline—has entirely lost its blue color. This should require about two hours, but the process may continue for three or four hours without injury to the analysis. The deposited nickel should be bright and coherent. Wash the deposited metal carefully but thoroughly by repeatedly pouring small portions of hot water upon it. Dry it in a water-bath, and then weigh it.

Care of the Apparatus.

Clean the Battery.—Pour away the acids used, and after washing the various parts of the apparatus, set them in place and fill them with water; leave the battery in this condition.

Clean the Platinum Dish.—Warm some Nitric acid in it, and afterward thoroughly wash it with water.

Note.

101. Give a careful consideration to the notes on pages 107, 108, and 109, but observe in connection with note 68 that Nickel deposits better in a solution made alkaline with Ammonia.

TWENTY-SIXTH EXERCISE—NICKEL.
(THE OXIDE METHOD.)

Data.

	Molecular Weight.	Per Cent.
Ni	57.9	14.714
O	16.	4.066
SO$_3$	79.9	20.305
(NH$_4$)$_2$	36.	9.148
O	16.	4.066
SO$_3$	79.9	20.305
6 H$_2$O	108.	27.395
	393.7	99.999
Ni	57.9	78.349
O	16.	21.651
	73.9	100.000

The Compound Tested.

The substance tested is the double Sulphate of Ammonia and Nickel, also called Ammonio-nickelous sulphate, $(NH_4)_2 SO_4 + Ni SO_4 + 6 H_2 O$.

Outline of the Process.

Precipitate the nickel as Nickelous hydrate ($Ni O_2 H_2$). Oxidize this compound to Nickelic hydrate ($Ni_2 O_6 H_6$) by means of Bromine water. Change the Nickelic hydrate by ignition to the Nickelous oxide ($Ni O$), and weigh it as such.

The Process.

The Weighing.—Weigh one gramme of Ammonio-nickelous sulphate, $(NH_4)_2 SO_4 + Ni SO_4 + 6 H_2 O$.

The Dissolving.—Dissolve the weighed salt in water.

The Precipitating.—To the solution add a slight ex, cess of Sodium hydroxide, and boil. The Nickel is precipitated as Nickelous hydroxide ($Ni\ O_2\ H_2$).

$$[Ni\ SO_4 + (NH_4)_2\ SO_4 + 6\ H_2\ O] + 4\ Na\ OH =$$
$$Ni\ O_2\ H_2 + 2\ Na_2\ SO_4 + 2\ NH_3 + 2\ H_2\ O + 6\ H_2\ O.$$

After boiling for some time *to expel all the Ammonia*, carry the solution to the hood and there add a little Bromine water and again boil. The Bromine oxidizes the light green Nickelous hydroxide into the black Nickelic hydroxide ($Ni_2\ H_6\ O_6$).[102]

$$2\ Ni\ O_2\ H_2 + 2\ Br + 2\ H_2\ O = Ni_2\ H_6\ O_6 + 2\ H\ Br.$$

After thorough boiling, allow the precipitate to subside.

The Filtration.—Decant the clear liquid upon a filter; boil the precipitate again with more water. After washing by decantation several times, transfer the precipitate to the filter and again wash thoroughly with boiling water.[103]

The Burning.—Dry; burn the filter first, then the precipitate; after heating strongly for a few minutes, weigh as Nickelous oxide ($Ni\ O$).

$$Ni_2\ H_6\ O_6\ \textit{heated} = 2\ Ni\ O + 3\ H_2\ O + O.$$

Notes.

102. Two advantages arise from the oxidizing of the Nickelous hydroxide into the Nickelic hydroxide : owing to the darker color of the latter, there is less danger that small particles of it will escape the attention of the analyst; again, owing to its more granular structure, it is easier washed.

103. Thorough washing is necessary to remove the excess of Sodium salts which adhere to the precipitate with great tenacity.

TWENTY-SEVENTH EXERCISE—NITROGEN.

(DISTILLATION METHOD.)

Data.

	Molecular Weight.	*Per Cent.*
N	14.	26.217
H_4	4.	7.491
Cl	35.4	66.292
	53.4	100.000
N_2	28.	6.326
$(H_4)_2$	8.	1.808
Pt	194.4	43.922
Cl_6	212.2	47.944
	442.6	100.000

The Compound Tested.

The substance tested is Ammonium chloride, Sal-ammoniac, (NH_4) Cl.

Outline of the Process.

Decompose the Ammonium salt by Sodium hydroxide in a retort or other suitable apparatus. Distil the mixture collecting the distillate in dilute Hydrochloric acid.

Evaporate to dryness the solution thus obtained and weigh the Ammonium chloride left.

The Process.

The Weighing.—Weigh one gramme of Ammonium chloride, (NH_4) Cl.

The Apparatus.—Place the weighed salt in a retort, the neck of which passes into a flask containing an excess of pure dilute hydrochloric acid. Suspend the retort in such a manner that the end of the neck will dip just beneath the surface of the liquid in the flask. Introduce into the retort, through the tubulature, sufficient water to dissolve the Ammonium chloride; add now a slight excess of Sodium hydroxide (Na OH).

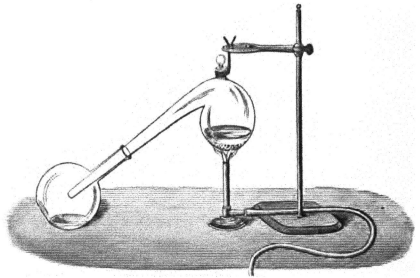

FIG. 34.—Arrangement of Apparatus for Distillation.

The Distillation. — Close the retort and gently heat it. The Sodium hydroxide reacts upon the Ammonium chloride according to the following equation :

$$(NH_4)\ Cl + Na\ OH = NH_3 + H_2\ O + Na\ Cl.$$

The Ammonia gas formed distils over and is absorbed by the Hydrochloric acid in the flask.

$$NH_3 + H\ Cl = (NH_4)\ Cl.$$

The Evaporation.—When all the Ammonia gas has passed over, remove the flask; wash the neck of the retort with distilled water, so that the washings pass into the flask. Transfer the solution of Ammonium chloride to a small weighed beaker and evaporate to dryness on a water-bath. Dry at 212° F. until the weight of the residue remains constant. From the weight of Ammonium chloride calculate the weight of Nitrogen.

Notes.

104. Ammonium compounds may be tested by means of solution of Platinic chloride (or Hydrochloroplatinic acid) and by a method similar in almost all respects to that employed with Potassium. (See description at page 153.)

The solution of the Ammonium salt is evaporated on a water bath with an excess of Platinic chloride ($Pt\,Cl_4$) (or Hydrochloroplatinic acid, $H_2\,Pt\,Cl_6$). The nitrogen is precipitated in the form of Ammonioplatinic chloride $(NH_4)_2\,Pt\,Cl_6$. This substance, after evaporation, etc., is weighed. This process is applicable with greatest advantage to those substances which contain Ammonium in the form of chloride.

When Ammonium salts other than the chloride are to be tested, if they can be changed to chloride by evaporation to dryness with hydrochloric acid, this should be done. The chloride so produced may then be dissolved in water and tested with Platinic chloride as just suggested.

This Platinum method may also be advantageously applied, as a confirmatory test, to the solution of Ammonium chloride obtained by the distillation method described in the foregoing Exercise.

13   K

TWENTY-EIGHTH EXERCISE—NITROGEN.
(PELOUZE'S METHOD MODIFIED.)

Data.

	Molecular Weight.	Per Cent.
K	39.	38.652
N	14.	13.875
O_3	47.9	47.473
	100.9	100.000
K_2	78.	38.652
O	16.	7.929
N_2	28.	13.875
O_5	79.8	39.544
	201.8	100.000

The Compound Tested.

This is Potassium nitrate (KNO_3). The process is applicable to any nitrate that yields Nitric acid, and no other oxidizing agent, by boiling with Hydrochloric acid in presence of Ferrous chloride.

Outline of the Process.

(*a*) Prepare a solution of Ferrous chloride containing a known amount of iron.

(*b*) To this *Ferrous* solution add the compound to be tested under such conditions as will force the nitrogen compound to exert its full oxidizing power upon the iron.

(*c*) By means of a standard solution of Stannous chloride estimate the amount of *Ferric* salt formed, and thence. by calculation (based on the reaction on page 149), determine the amount of the Nitrogen compound that accomplished the oxidation.

(*c'*) Instead of the treatment described in (*c*), the analyst may determine by means of standard solutions the amount of iron *unoxidized.* Since the amount of Iron first taken is accurately known, the amount oxidized may be learned by subtraction; thence by calculation, as suggested in (*c*), the amount of Nitrogen compound may be estimated.

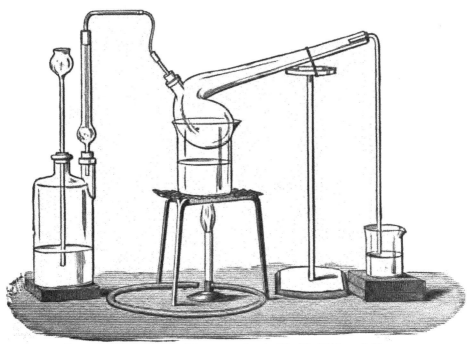

FIG. 35.—Arrangement of Apparatus for the Modified Pelouze's Process.

The Apparatus Used.

First. A generator of Carbon dioxide.[106]

This consists of a large Woulff's bottle having two necks. One neck is for the introduction of hydrochloric acid to liberate the Carbon dioxide from marble ($Ca\ CO_3$). A relatively large amount of marble should be used. (See page 79.)

$$Ca\ CO_3 + 2\ H\ Cl = CO_2 + Ca\ Cl_2 + H_2\ O.$$

The other neck of the Woulff's bottle is supplied with a washing-tube provided with moist fragments of sponge.

Second. A tubulated retort.

At the tubulation there is a glass tube projecting about one inch into the retort; this tube is connected, *air-tight*, with the supply of Carbon dioxide. The neck of the retort should incline upward.

Third. A small beaker containing water.[107] A glass tube, connected air-tight with the neck of the retort, dips into the water in this beaker.

The Process.

The Weighing.—Weigh accurately about 1.5 grammes of fine piano-forte wire. This is to be used in preparation of the Ferrous solution.

Weigh accurately, in a short test-tube, about 400 milligrammes of pure Potassium nitrate This is the substance to be tested.

Carefully preserve the substances weighed until the apparatus is ready for them.

The Ferrous Solution.—Adjust the apparatus already described, as shown in Fig. 35.

Pour into the retort about 25 to 30 cubic centimetres of pure concentrated hydrochloric acid; connect the retort with the Carbon dioxide generator; pass a steady flow of the gas into the retort to completely expel the air. Introduce the weighed iron wire into the retort through the neck, and gently warm the whole until the iron wire is completely dissolved.

$$Fe + 2\,H\,Cl = Fe\,Cl_2 + H_2.$$

Introduction of the Nitrate.—At this stage increase the flow of Carbon dioxide and introduce the nitrate—contained in the small test-tube—through the neck of the retort. Place a water-bath under the retort and digest the contents of the retort for about a quarter of an hour.

$$6\,Fe\,Cl_2 + 2\,K\,NO_3 + 8\,H\,Cl = 3\,Fe_2\,Cl_6 + N_2\,O_2 + 2\,K\,Cl + 4\,H_2\,O.$$

Withdraw the water-bath and raise the heat to boiling. Keep the retort at this temperature until the Nitrogen dioxide ($N_2\,O_2$) is completely expelled; when this point is reached, the solution in the retort, though of a deep yellow or red color, is clear.

The Titration.—Remove the iron solution to a casserole and estimate the amount of iron oxidized. Accomplish this by means of a standard solution of Stannous chloride, and according to the directions given at pages 122–128, and in paragraph (*c*) on page 147. From the amount of Ferric compound found estimate the amount of Nitrogen or Nitric acid, as required, being guided by the equation given above.

Notes.

105. If the Nitrogen compound to be tested is in the form of a liquid, it may be introduced into the retort, through the tubulature, by means of a funnel.

106. The purpose of Carbon dioxide is to replace the air in the apparatus. If the oxygen of the air is allowed in presence of hot Ferrous chloride, it produces Ferric chloride, and thus leads us to infer the presence of too large an amount of Nitric acid.

13*

107. The purpose of the water-beaker is twofold. It prevents the entrance of the oxygen of the air at that end of the apparatus. It also absorbs the hydrochloric acid gas expelled by the boiling.

108. After the Nitrogen compound has oxidized the iron and the iron solution has been cooled in the stream of Carbon dioxide, the amount of iron *unoxidized* may be estimated.

For this purpose Penny's process may be employed. This involves a standard solution of Potassium dichromate of such strength that one cubic centimetre corresponds to 10 milligrammes of metallic iron. The solution of Dichromate is drawn from a burette, little by little, into the Iron solution. The end of the reaction is recognized as the point at which the Iron solution ceases to give a blue precipitate, when a drop is added to a drop of a solution of Potassium ferricyanide upon a white porcelain plate. Of course the Ferricyanide used must be free from Ferro-cyanide.

TWENTY-NINTH EXERCISE—PHOSPHORUS.

Data.

	Molecular Weight.	*Per Cent.*
H	1.	.280
Na$_2$	46.	12.867
P	31.	8.671
O$_4$	63.9	17.874
12 H$_2$O	215.6	60.308
	357.5	100.000
H$_2$	2.	.280
O	16.	2.239
2 Na$_2$	92.	12.873
O$_2$	31.9	4.463
P$_2$	61.9	8.661
O$_5$	79.8	11.166
24 H$_2$O	431.1	60.319
	714.7	100.001
Mg$_2$	47.9	21.625
O$_2$	31.9	14.402
P$_2$	61.9	27.946
O$_5$	79.8	36.027
	221.5	100.000

The Compound Tested.

This is crystallized Hydro-disodium phosphate (H Na$_2$ PO$_4$ + 12 H$_2$ O).

Outline of the Process.

Precipitate the Phosphorus as Ammonio-magnesium phosphate.

By ignition change the precipitated compound into Magnesium pyro-phosphate, in which form it is to be weighed.

The Process.

The Weighing.—Weigh one gramme of the salt to be tested, in this case, Hydro-disodium phosphate.

The Dissolving. — Dissolve the salt in water; add a few cubic centimetres of a solution of Ammonium chloride.

The Precipitation.—To the clear solution add a measured amount of "Magnesia solution." (See p. 59.) The solution should be added from a burette, at the rate of about one drop per second. The beaker must be shaken all the time (not stirred with a rod). The amount used must represent a slight excess over the amount calculated. Allow the whole to stand for a few minutes, and then add an amount of Ammonium hydroxide equal in bulk to one-third the volume of the solution. Allow the whole to stand for from twelve to twenty-four hours.

The Filtration.—Pass the clear liquid through a filter; then, after washing by decantation and with water made alkaline with a small amount of Ammonium hydroxide, transfer the precipitate to the paper.

The Burning.—Dry the precipitate; separate the filter-paper and burn it in a platinum crucible; add the precipitate and ignite it, *gently at first*, afterward with the blast-lamp.

$$2 \, (NH_4 \, Mg \, PO_4) \; heated = Mg_2 \, P_2 \, O_7 + 2 \, NH_3 + H_2 O.$$

The Calculation. — Weigh as Magnesium pyro-phosphate ($Mg_2 \, P_2 \, O_7$). From the weight of this substance calculate the weight of Phosphorus.*

* Compare Exercise on Magnesium, pp. 134, 135.

THIRTIETH EXERCISE—POTASSIUM.

Data.

	Molecular Weight.	Per Cent.
K	39.	52.419
Cl	35.4	47.581
	74.4	100.000
K_2	78.	16.096
Cl_2	70.7	14.589
Pt	194.4	40.116
Cl_4	141.5	29.199
	484.6	100.000

The Compound Tested.

This is Potassium chloride (K Cl).

Outline of the Process.

Precipitate the Potassium as Potassio-platinic chloride (K_2 Pt Cl_6), and weigh it as such.

If the Potassium is not in the form of Potassium chloride, change it to this compound before precipitation.

The Reagents.

Platinum solution.—This should contain 1 gm. of metallic Platinum in 10 c. c. of solution.

Ammonium chloride solution. — Place in a bottle 500 c. c. water ; add 100 gms. Ammonium chloride ; shake until dissolved. Pulverize 5 or 10 gms. of Potassio platinic chloride; add to the solution of Ammonium chloride; shake at intervals for six or eight hours ; leave it over night ; filter the clear liquid for use. Keep the residue for preparation of a fresh supply.

Sodium chloride solution. — Dissolve 20 gms. pure Sodium chloride in water, and dilute to 1000 c. c.

The Process.

The Weighing. — Weigh 1 gm. of Potassium chloride.

The Dissolving.—Dissolve the weighed salt in about 10 c. c. of water; wash the solution into a 200 c. c. flask; dilute to the mark. From this solution draw, for each test, 50 c. c. (representing 250 milligrammes of K Cl) and transfer this portion to a porcelain dish.

The Precipitation.—To the solution add 5 c. c. of solution of Sodium chloride; then add an excess of Platinum solution (4 c. c.).[109]

$$2 \text{ K Cl} + \text{Pt Cl}_4 = \text{K}_2 \text{ Pt Cl}_6.$$
$$2 \text{ K Cl} + \text{H}_2 \text{ Pt Cl}_6 = \text{K}_2 \text{ Pt Cl}_6 + 2 \text{ H Cl}.$$

Evaporate the solution to dryness on the water bath; (do not have the water strongly boiling).

To the residue add 10 c. c. water, and then strong alcohol.

FIG. 36.—Water-bath Arrangement for Evaporating the Solution to be Tested for Potassium.

The Filtration, etc. —Prepare a balanced filter. Wash the precipitate thoroughly with alcohol by decantation upon the filter. Continue the washing even after the filtrate is colorless. Then pass through the precipitate 10 c. c. of the Ammonium chloride solution prepared as directed; (these 10 c. c. contain the bulk of the impurities and are to be thrown away.) Next add fresh portions of 10 c. c. each of Ammonium chloride solution; repeat this washing five or six times. Then wash the filter thoroughly with pure alcohol. Dry the precipitate at 266° F. (130° C.), and weigh as Potassio-platinic chloride ($K_2 Pt Cl_6$). From the weight of this substance calculate the weight of Potassium.

Notes.

109. Owing to the high cost of Platinum, its solutions are usually made of a known strength; that is, they are made to contain a known number of milligrammes of Platinum per cubic centimetre of the solution.

110. The amount of Platinum that is actually required to unite with the Potassium in the amount of Potassium chloride taken, may be calculated by use of the following proportion :

Twice the molecular weight of K Cl,

is to the atomic weight of Platinum,

as the gross weight of K Cl,

is to the gross weight of Platinum required.

It is of great advantage to calculate and record, once for all, the amount of Potassium that each cubic centimetre of a given Platinum solution should precipitate.

111. Potassio-platinic chloride is slightly soluble in cold water, more readily in hot water. It is nearly insoluble in absolute alcohol and but sparingly soluble in Spirit of wine.

THIRTY-FIRST EXERCISE—SILICON.

Data.

	Molecular Weight.	Per Cent.
K_2	78.	14.011
O	16.	2.874
Al_2	54.	9.700
O_3	47.9	8.604
6 Si	169.2	30.393
6 O_2	191.6	34.417
	556.7	99.999
Si	28.2	46.922
O_2	31.9	53.078
	60.1	100.000

The Compound Tested.

This is the variety of felspar called orthoclase (K_2 O, Al_2 O_3, 6 Si O_2). Certain specimens of felspar may differ somewhat in composition from that represented in the above table.

Outline of the Process.

(*a*) Fuse the powdered mineral with an alkaline carbonate (K_2 CO_3); this changes the insoluble mineral silicate into a soluble alkaline silicate.

(*b*) Dissolve the alkaline silicate in water; add hydrochloric acid to liberate the Silicic acid.

(*c*) Evaporate the mass to dryness to produce Silicic anhydride.

(*d*) Digest the dry residue in water and hydrochloric

acid; this leaves Silicic anhydride ($Si\,O_2$), which is the substance to be weighed.

The Process.

The Weighing.—Weigh in a platinum crucible 500 milligrammes of the *very finely powdered* Silicate.

The Fusing.—Into the crucible introduce about 2.5 grammes of pure Potassium or Sodium carbonate, and after thorough mixing, by stirring carefully with a platinum wire, heat the mixture with the blast-lamp.[113] At first warm gently to avoid loss through the escape of Carbon dioxide, for when the heat is too great this is apt to throw out of the crucible small portions of the undecomposed silicate. Next heat to fusion and maintain the mass in that condition until the contents of the crucible are in the form of a quiet and perfectly transparent liquid. Finally allow the whole to cool.

$$(K_2\,O,\ Al_2\,O_3,\ 6\,Si\,O_2) + 12\,K_2\,CO_3 = 6\,K_4\,Si\,O_4 + K_2\,Al_2\,O_4 + 12\,CO_2.$$

Separating the Silica.—Place crucible and contents in a casserole provided with a glass cover. Cover the crucible with hot water and then bring the liquid to boiling; this is for the purpose of dissolving the fused mass so far as may be practicable. Next add pure concentrated Hydrochloric acid, taking care to avoid loss through effervescence. When the reaction seems complete, remove the now clean crucible, washing its surface into the casserole by means of a wash-bottle.

$$K_4\,Si\,O_4 + 4\,H\,Cl = H_4\,Si\,O_4 + 4\,K\,Cl.$$

Place the casserole, with its contents, on a water-bath[115] and evaporate to complete dryness. When the mass is dry allow it to cool; then add distilled water and about a cubic centimetre of pure concentrated hydrochloric acid;[116] then gently warm the mixture again.

$$H_4\,Si\,O_4\ (heated) = Si\,O_2 + 2\,H_2\,O.$$

14

The Filtering.—Filter the mass and wash the precipitate thoroughly with boiling water.[117]   Finally dry the precipitate, ignite it, and weigh it as Silicic oxide, also called Silica (Si O_2).

Notes.

112. The change of the silicates to the soluble form depends upon the fact that though all silicates are insoluble when the Silica predominates, the alkaline silicates are soluble when the alkali metal predominates.

113. Great care must be taken in using platinum vessels.   When heated, they should rest on *platinum triangles;* there should not be heated in them any substances likely to injure them.   (Read carefully note 35, p. 74.)

114. It is always better to thoroughly dissolve the fused mass in water before adding hydrochloric acid, otherwise the action of the acid—by decomposing the alkaline Silicate upon the *outside* of the mass—coats the whole over with insoluble Silicic acid, before the *inside* of the mass is acted upon.

115. Drying upon a water-bath is necessary; upon the application of strong heat to the gelatinous precipitate of Silicic acid the steam formed within the lumps bursts them, and thus occasions loss.

116. Hydrochloric acid is added for the purpose of dissolving metallic oxides, and in general all substances other than Silica.

117. Thorough washing of the Silica is necessary, since it retains, with great tenacity, portions of the alkaline salts produced by the experiment.

THIRTY-SECOND EXERCISE—SILVER.

Data.

	Molecular Weight.	Per Cent.
Silver coin of the United States,		
Ag		90.00
Cu		10.00
		100.00
Ag	107.7	75.262
Cl	35.4	24.738
	143.1	100.000

The Compound Tested.

This is a silver coin of the United States; it should contain 90 per cent. silver and 10 per cent. copper.

Outline of the Process.

Dissolve the coin in Nitric acid. Precipitate the Silver as Silver chloride (Ag Cl), and weigh it in that form.

The Process.

The Weighing.—Weigh a clean silver ten-cent piece. (Its weight should be approximately 2.488 grammes.)

The Dissolving.—Place the coin in a covered beaker, add dilute Nitric acid, and warm gently until the alloy is completely dissolved.

$$2\,Cu\,Ag + 8\,H\,NO_3 = 2\,Cu\,(NO_3)_2 + 2\,Ag\,NO_3 + N_2O_2 + 4\,H_2O.$$

Dilute the solution with water so as to bring it to the volume of 500 cubic centimetres; agitate the solution so that it may be homogeneous. Take three separate fifths of the solution and test each according to the following description.

The Precipitation.—To the portion of solution under examination add a slight excess of pure dilute hydrochloric acid. Warm the mixture and stir it vigorously until —after allowing the precipitate to subside—the supernatant liquid is clear, or very nearly so.

$$Ag\ NO_3 + H\ Cl = Ag\ Cl + H\ NO_3.$$

The Filtering.—Decant the clear liquid upon a filter, and after several washings by decantation, place the bulk of the precipitate on the filter; wash with hot water until the washings are neutral.

The Burning.—Dry the precipitate, and when dry, remove it as completely as possible from the paper. Burn the paper first (in a porcelain crucible); allow the ash to cool, then add a drop of pure Nitric acid, and, after warming, a drop of pure Hydrochloric acid; carefully evaporate the liquid to dryness and heat the residue to fusion. After allowing the crucible to cool, add the precipitate; heat until the Silver chloride begins to fuse. Cool and weigh.

The Calculation. — From the weight of the Silver chloride calculate the weight of silver present.

Note.

118. This process must be performed away from direct sunlight. Under the influence of strong light the Silver chloride is decomposed, with loss of chlorine. (See notes under chlorine, pages 90 and 91.)

THIRTY-THIRD EXERCISE—SULPHUR.
(METHOD FOR SULPHATES.)

Data.

	Molecular Weight.	Per Cent.
Cu	63.2	25.392
O	16.	6.428
S	32.	12.857
O_3	47.9	19.245
$5 H_2O$	89.8	36.079
	248.9	100.001
Ba	136.8	58.788
O	16.	6.876
S	32.	13.752
O_3	47.9	20.584
	232.7	100.000

The Compound Tested.

This is Cupric sulphate, Blue vitriol ($Cu\,SO_4 + 5\,H_2O$).

Outline of the Process.

Precipitate the Sulphur in the salt as Barium sulphate ($Ba\,SO_4$), and weigh it in that form.

(Primarily, this process is applicable only to Sulphuric acid and Sulphates, but it may be extended to other compounds after the sulphur in them has been oxidized to one of these forms.)

The Process.

The Weighing.—Weigh one gramme of Cupric sulphate ($Cu\,SO_4 + 5\,H_2O$).

The Dissolving.—Place the salt in a beaker and dis-

14 * L

solve it in hot water. Add a few drops of pure concentrated hydrochloric acid.[119]

The Precipitating.—To the solution add a slight excess of a solution of Barium chloride (Ba Cl$_2$). Boil for a few minutes; then allow the precipitate to subside.

$$Cu\ SO_4 + Ba\ Cl_2 = Ba\ SO_4 + Cu\ Cl_2.$$

The Filtering.—Pass the clear liquid through a filter; wash the precipitate thoroughly by decantation; transfer the precipitate to the filter.

The Burning, etc.—Dry the precipitate and then remove it as completely as possible from the filter-paper.[121] Burn the filter first. To the ash add a drop of Sulphuric acid; carefully evaporate the product to dryness and ignite it. Allow the residue to cool; then add the reserved precipitate. After heating strongly for a few minutes, place the crucible in a desiccator to cool.

The Calculation.—From the weight of the Barium sulphate calculate the weight of sulphur.

Notes.

119. The Hydrochloric acid is added for the purpose of holding in solution sulphates other than Barium sulphate. This acid must not, however, be present in too large an excess, as the strong acid precipitates Barium chloride.

120. The particles of the precipitate—Barium sulphate—are exceedingly fine; sometimes they pass through the pores of the filter-paper. Thorough boiling overcomes, to a large extent, this difficulty.

121. The carbon and hydrogen of the filter-paper reduce the Barium sulphate to Barium sulphide.

$$Ba\ SO_4 + C_6\ H_{10}\ O_5 = Ba\ S + 5\ H_2\ O + 2\ CO_2 + 4\ C.$$

Adding Sulphuric acid to the filter-ash changes this Barium sulphide back to the sulphate.

$$Ba\ S + H_2\ SO_4 = Ba\ SO_4 + H_2\ S.$$

THIRTY-FOURTH EXERCISE—TIN.
(VOLUMETRIC METHOD.)

Data.

	Molecular Weight.	Per Cent.
Sn	117.7	52.474
Cl$_2$	70.7	31.520
2 H$_2$O	35.9	16.005
	224.3	99.999

The Compound Tested.

This is crystallized Stannous chloride, known in commerce as Tin crystals (Sn Cl$_2$ + 2 H$_2$ O).

Outline of the Process.

Oxidize the tin of the Stannous chloride to the Stannic form by means of a standard solution of Potassium dichromate (K$_2$ Cr$_2$ O$_7$). As an indicator for the end of the reaction, use a solution of Potassium iodide with starch.

The Standard Solutions.

1. Standard solution of Potassium dichromate.[122]

Dissolve 10.912 grammes of the pure dry salt in a small amount of hot water, and dilute the solution to the volume of 500 cubic centimetres. The strength of this solution is thus adjusted so that 1 cubic centimetre of it shall correspond to

50 milligrammes of S̀n Cl$_2$ + 2 H$_2$ O
equivalent to
26.24 milligrammes of metallic Tin.

2. A solution of Potassium iodide (KI), with starch.

Boil with constant stirring—

> 2 grammes of powdered starch,
> 6 grammes of Potassium iodide,
> in 200 cubic centimetres of water.

Allow the whole to cool.

The Process.

The Weighing.—Weigh about 10 grammes of the Stannous chloride (Sn Cl_2 + 2 H_2 O).

The Dissolving.—Dissolve the weighed salt in water containing one-half its bulk of pure concentrated hydrochloric acid—heating, if necessary.[87] When the salt is dissolved, dilute the solution to the volume of 500 cubic centimetres.

The Titration.—Measure off 100 cubic centimetres of the solution, equivalent to 2 grammes of the original Tin salt. Place this portion in a casserole, and add 5 cubic centimetres of pure concentrated hydrochloric acid and 5 cubic centimetres of the Starch liquor.

Now cautiously draw in, from a burette, the standard Potassic dichromate solution, until a permanent blue color of Iodine-with-starch is produced.

Repeat the process with three different fifths of the Stannous chloride solution.

The Calculation. — From the average number of cubic centimetres of Potassium dichromate solution added, calculate the exact amount of Tin in the 2 grammes of the Stannous chloride used; from this last weight calculate the percentage amount.

Note.

122. The action of the Potassium dichromate upon the Stannous chloride is indicated by the following equation :

$$3 \, Sn \, Cl_2 + K_2 \, Cr_2 \, O_7 + 14 \, H \, Cl = 3 \, Sn \, Cl_4 + 2 \, K \, Cl + Cr_2 \, Cl_6 + 7 \, H_2 \, O.$$

Hence the amount of Potassium dichromate necessary for the standard solution may be obtained by the following proportion :

Molecular Weight of $3 \, (Sn \, Cl_2 + 2 \, H_2 \, O)$	:	Molecular Weight of $K_2 \, Cr_2 \, O_7$	::	Grammes of $Sn \, Cl_2 + 2 \, H_2 \, O$	:	Grammes of $K_2 \, Cr_2 \, O_7.$
672.9	:	293.7	::	$\left\{\begin{array}{c} 500 \\ \times \\ 50 \text{ milligrammes} \\ = 25.000 \text{ Gms.} \end{array}\right\}$	:	10.912 Gms.

THIRTY-FIFTH EXERCISE—ZINC.

Data.

	Molecular Weight.	Per Cent.
Zn	64.9	22.653
O	16.0	5.585
S	32.0	11.169
O_3	47.9	16.719
7 H_2O	125.7	43.874
	286.5	100.000
Zn	64.9	80.222
O	16.0	19.778
	80.9	100.000

The Compound Tested.

This is crystallized Zinc sulphate, White vitriol ($Zn SO_4 + 7 H_2 O$).

Outline of the Process.

Precipitate the zinc as a Carbonate; change it into the oxide ($Zn O$) by ignition; weigh it in the latter form.

The Process.

The Weighing.—Weigh 1 gramme of crystallized Zinc sulphate ($Zn SO_4 + 7 H_2 O$).

The Dissolving.—Place the salt in a casserole and dissolve it in water. Heat the solution to boiling.

The Precipitating.—To the hot solution carefully add a slight excess[124] of Sodium carbonate ($Na_2 CO_3$).

After boiling for a few minutes, allow the precipitate to subside.

$$Zn\ SO_4 + Na_2\ CO_3 = Zn\ CO_3 + Na_2\ SO_4.$$

The Filtering.—Pass the clear liquid through a filter; add more water to the precipitate, then boil and decant as before. After washing by decantation three times, place the precipitate upon the paper and wash thoroughly with boiling water.

The Burning.—Dry the precipitate, then remove it as completely as possible from the paper. Burn the paper first[125] in a porcelain crucible, then add the rest of the precipitate, and strongly ignite for some time.

$$Zn\ CO_3\ \textit{heated} = Zn\ O + CO_2.$$

Weigh the Zinc oxide formed, and from its weight calculate the weight of zinc.

Notes.

123. From zinc solutions Sodium carbonate precipitates a basic Zinc carbonate with liberation of Carbonic acid; the precipitate is of varying constitution, but the equations in the text have been constructed with reference to the normal Carbonate.

124. When a neutral solution of zinc is precipitated by Sodium carbonate, a portion of the zinc is held in solution by the Carbonic acid liberated. By thorough boiling, however, the zinc is completely precipitated.

125. The precipitate must be completely removed from the paper, because (through the action of the carbon and hydrogen of the filter-paper) reduction and volatilization of the metallic zinc take place. These sources of loss may be avoided by saturating the paper before burning with a solution of Ammonium nitrate, $(NH_4)NO_3$.

APPENDIX.

Supplies Needed for Quantitative Analysis.

I. CLASSIFIED LISTS.

Chemically Pure Substances to be Analyzed.

1. Ammonio-aluminic sulphate (ammonia alum).
2. Potassio-antimonylic tartrate (tartar-emetic).
3. Arsenious oxide (white arsenic).
4. Baric chloride (chloride of barium).
5. Bismuthyl nitrate (nitrate of bismuth).
6. Potassium bromide (bromide of potassium).
7. Calcium carbonate (Iceland spar).
8. Sodium chloride (common salt).
9. Potassium dichromate (bichrome)
10. Cupric sulphate (sulphate of copper).
11. Ammonio-ferric sulphate (iron alum).
12. Ammonio-ferrous sulphate (double sulphate of iron and ammonia).
13. Lead nitrate (nitrate of lead).
14. Lead sulphide (galena).
15. Magnesium sulphate (Epsom salt).
16. Mercuric chloride (corrosive sublimate).
17. Ammonio - nickelous sulphate (double sulphate of nickel and ammonia).
18. Ammonium chloride (sal-ammoniac).
19. Potassium nitrate (saltpetre).
20. Hydro - disodium phosphate (phosphate of soda).
21. Potassium chloride (chloride of potassium).
22. Felspar.
23. Silver coin.
24. Stannous chloride (tin crystals).
25. Zinc sulphate (sulphate of zinc).

Chemical Reagents.

26. Acid, acetic.
27. —— hydrobromic.
28. —— hydrochloric, commercial.
29. —— —— pure.
30. Acid, nitric, commercial.
31. —— —— pure.
32. —— sulphuric, commercial.
33. —— —— pure.

34. Acid, sulphydric.
35. ——— tartaric.
36. Alcohol, ethyl.
37. Ammonium acetate.
38. ——— carbonate.
39. ——— chloride.
40. ——— nitrate.
41. ——— oxalate.
42. Silver nitrate.
43. Bromine.
44. ——— water.
45. Calcium chloride.
46. Carbon disulphide.
47. Iodine.
48. Magnesium chloride.
49. Marble.

50. Mercuric chloride.
51. Platinic chloride.
52. Potassium carbonate.
53. ——— chlorate.
54. ——— chromate.
55. ——— ferricyanide.
56. ——— ferrocyanide.
57. ——— iodide.
58. ——— permanganate.
59. Sodium carbonate.
60. ——— hydrate.
61. Starch.
62. Water, distilled.
63. Wire, iron.
64. ——— piano-forte.
65. Zinc.

General Stock of Apparatus.

66. Apparatus, Johnson's, for Carbon dioxide.
67. ——— for determination of Nitrogen in Nitrates.
68. ——— Scheibler's, for Carbon dioxide.
69. ——— sulphuretted hydrogen.
70. Battery, galvanic.
71. Bellows.
72. Blast-lamp.
73. Boat, porcelain.
74. Bottles, glass stoppered.
75. Burettes and fittings.
76. Corks, rubber.
77. Cotton.
78. Crucibles, platinum.
79. Desiccators.
80. Dishes, evaporating.
81. Files.
82. Filter, cutter.

83. Filter-pump.
84. ——— paper.
85. Flasks, graduated.   [tion.
86. Glass tube, Bohemian combus·
87. Graduates.
88. Hydrometers.
89. Labels.
90. Lamp.
91. ———, blast.
92. Mortar, agate.
93. ——— wedgewood.
94. Oven, drying.
95. Paper, filter.
96. Platinum, cones.
97. ——— crucibles.
98. ——— triangles.
99. Retorts.
100. Thermometer.
101. Thistle tubes.
102. Woulff's bottle.

Articles for each Student's Desk.

103. Beakers.
104. Brushes, camel's-hair.
105. Casserole.

106. Crucibles (porcelain).
107. Feather.
108. Filters.

15

109. Funnels.
110. Gauze (iron).
111. Glass cover.
112. ——— rods.
113. Lamp, Bunsen.
114. Paper, filter.
115. ——— glazed.

116. Paper, litmus.
117. Triangles, iron.
118. Tripod or lamp-stand.
119. Tweezers.
120. Wash-bottle.
121. Watch-glasses.
122. Weights.

II. ALPHABETICAL DESCRIPTIVE LIST OF ALL SUPPLIES NEEDED.

Acid, hydrobromic, **H Br.**

——hydrochloric, *Pure,* **H Cl.**

——— nitric, *pure,* **H NO$_3$.**

——— nitric, *fuming,* **N NO$_3$.** (See p. 56.)

——— tartaric, **H$_4$ O$_4$ $\overline{\text{T}}$.** (See p. 51.)

——— hydrosulphuric **H$_2$ S.** (See p. 55.)

——— sulphuric, *pure.* **H$_2$ SO$_4$.**

——— sulphuric, *dilute,* **H$_2$ SO$_4$ + Aq.**

Add one part of Sulphuric acid by measure to five parts of distilled water by measure.

Alcohol, ethyl (ordinary alcohol), **(C$_2$ H$_5$) OH.**

Ammonia alum (see Ammonio-aluminic sulphate), **(NH$_4$)$_2$ SO$_4$ + Al$_2$ (SO$_4$)$_3$ + 24 H$_2$ O.**

Ammonium acetate (Acetate of ammonia), **NH$_4$ O A** or **NH$_4$ O C$_2$ H$_3$ O.** (See p. 133.)

——— carbonate, **(NH$_4$)$_2$ CO$_3$.**

Dissolve 1 part of commercial Sesquicarbonate of ammonia in a mixture of 3 parts of water and 1 part of Ammonium hydroxide — all by weight.

——— chloride (Salammoniac), **NH$_4$ Cl.**

Dissolve 1 part of the dry salt in the form of powder in 8 parts of water.

——— hydrate (Ammonia), **NH$_4$ OH.**

Dilute the commercial concentrated Ammonia (that which has a specific gravity of .900) with twice the volume of water.

——— nitrate, **NH$_4$ NO$_3$.**

——— oxalate, **(NH$_4$)$_2$ O$_2$ C$_2$ O$_2$.**

Ammonio-aluminic sulphate (Ammonia alum). (See p. 47.) **(NH$_4$)$_2$ SO$_4$ + Al$_2$ (SO$_4$)$_3$ + 24 H$_2$ O.**

Ammonio-ferrous sulphate (double sulphate of iron and ammonia), $(NH_4)_2$ $SO_4 + Fe\ SO_4 + 6\ H_2\ O$.

Ammonio-ferric sulphate (Ferric alum, Iron alum), $(NH_4)_2\ SO_4 + Fe_2$ $(SO_4)_3 + 24\ H_2\ O$. (See p. 48.)

Ammonio-nickelous sulphate (double sulphate of nickel and ammonia), $(NH_4)_2\ SO_4 + Ni\ SO_4 + 6\ H_2\ O$.

Apparatus for determination of Nitrogen in Nitrates by modification of Pelouze's method. (Described at p. 147.)

———— for determination of Carbon dioxide.
Johnson's form, described at p. 77.
Scheibler's form, described at p. 82.

———— for preparation of Sulphuretted hydrogen.
(Described at p. 55.)

Argentic nitrate (Nitrate of silver), **Ag NO₃.**

Ordinarily the crystallized salt, as occurring in commerce, is used for making the solution.

The metallic silver obtained from refining of silver residues (see below) may be used.

Dissolve 100 grammes of metallic silver in

   125 c. c. Nitric acid, and
   125 c. c. water.

Heat the mixture; when the metal is dissolved evaporate the solution on a water-bath until it becomes a mass of crystals.

Silver residues. It is desirable to have in the laboratory a large bottle in which to put all waste solutions containing silver, and, indeed, all such solids as chloride, bromide, and iodide of silver. As the waste accumulates make an occasional addition of hydrochloric acid to the mixture. At a convenient time filter, wash, dry and weigh the silver precipitate; then reduce it to metallic silver. Accomplish this reduction as follows: Add to the dried precipitate, mostly chloride, about twice its weight of Sodium carbonate. Transfer the mixture to a large Hessian crucible, and then heat the whole in a furnace until the mass is in limpid and quiet fusion. Allow the crucible to cool where it will not be disturbed. When it is cool, break off the base of the crucible and the slag; the slag should be free from detached globules of silver, but a mass of the metal—proportional to the weight of chloride used—should be at the bottom of the pot. Hammer the button on a clean anvil so as to break off the adhering slag, then boil it in water so as to remove such matters as remain in the cavity usually found at the top of the button. Finally dissolve the button of metal as already described.

Arsenious oxide (White arsenic), **As$_2$ O$_3$.**

Barium chloride,**Ba Cl$_2$ + 2 H 2 O.**

Balance.

> The principles upon which the accuracy and sensibility of the Balance depend are discussed at pp. 26 to 34.

Balloon (of rubber).

> It is convenient to have extra balloons for Scheibler's apparatus. (See pp. 82 to 86.)

(Barometer.

> The only determination in this book which gives occasion for the use of the barometer is that of Carbon dioxide by Scheibler's method; but even here it is not absolutely indispensable.)

FIG. 37.—Water-baths.

Baths.   (See Ovens.)

> The baths shown in Fig. 37 are very useful for evaporation or solutions from open beakers and other vessels.   They are simply open copper kettles supported on three legs and containing a relatively large supply of water.   At the top may be rings of various sizes to accommodate various vessels.

Battery, galvanic (Grove's).

> (See general account at pp. 107, 108, 109.)

Beakers.

There are two principal styles: the tall or French form and the wide or Griffin form. The latter are generally preferred for quantitative work.

Bellows.

The foot-bellows should be made double, that is, they should consist of an upper and a lower bellows. The up-and-down motion of

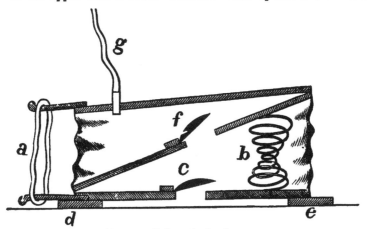

Fig. 38.—Bellows in Section.

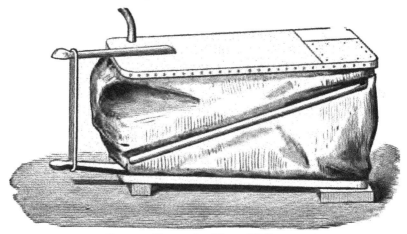

Fig. 39.—Perspective View of Bellows.

the foot forces air from the lower bellows into the upper. The upper one furnishes a steady supply of air, the continuous pressure being afforded by the rubber band a. A block at e and two strips near c and f prevent the bellows interfering with the action of the valves cf; the spring b opens the lower bellows automatically.

Bismuthyl nitrate (Subnitrate of bismuth), **Bi O NO$_8$**.

Blast-lamp. (See Lamp.)

15 *

Boat, porcelain.

This should be of such size as to easily slip into the Bohemian combustion-tube used.

Borax. (See Sodium tetra-borate.)

Bottles, glass-stoppered (for standard solutions).

Each student should have three or four bottles of about the capacity of one quart each.

Bromine.

Bromine water.

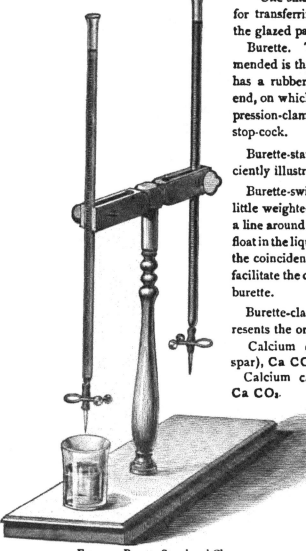

FIG. 40.—Burette Stand and Clamp.

Brush, camel's-hair.

One small one is convenient for transferring precipitates from the glazed paper to the crucible.

Burette. The variety recommended is that called Mohr's. It has a rubber joint at the lower end, on which is attached a compression-clamp that serves as a stop-cock.

Burette-stand. This is sufficiently illustrated by the figure.

Burette-swimmer. This is a little weighted bulb marked with a line around it. It is intended to float in the liquid in the burette, and the coincidences of its belt-line to facilitate the correct reading of the burette.

Burette-clamp. The figure represents the ordinary form.

Calcium carbonate (Iceland spar), $Ca\ CO_3$.

Calcium carbonate (Marble), $Ca\ CO_3$.

Calcium chloride, $Ca\ Cl_2$.

(See p. 80.) Carbon disulphide (Bisulphide of carbon), CS_2.

Casserole.

One of eight ounce capacity.

Corks.

Cotton.

Crucibles.

Those of porcelain are generally used, the ordinary size being those of $1\frac{1}{2}$ inch diameter at top.   Platinum are preferable.   (See page 20.)

Cupric sulphate (sulphate of copper), $\textbf{Cu SO}_4 + 5\ \textbf{H}_2\ \textbf{O}$.

FIG. 41.—Desiccators.

Desiccator.

A vessel in which a crucible may be cooled in dry air.   The drying material generally used is oil of vitriol.   The figures show different forms.

Dishes, evaporating.

Those made by the Royal Berlin and the Royal Dresden factories are of the highest grade.

Feather.

A banker's quill has usually a sufficient amount of the feathery part to fit it for use in removing the last portions of precipitates from beakers to the filters.

Felspar.   (See Potassio-aluminic silicate.)

Ferrous sulphide (Sulphide of Iron), $\textbf{Fe S}$.   (See p. 55.)

File.

A three-sided one for cutting glass tubes.

Filter-cutter and mallet.

Filter-paper is easiest cut by use of a cylindrical steel cutter.   It is convenient to have two sizes, one $4\frac{1}{2}$ inches in diameter, the other $3\frac{1}{2}$ inches in diameter.   (See Fig. 42.)

Filter-paper.

Some account of this paper is given at p. 15.

To determine the average amount of mineral ashes afforded by a given paper, select five clean filters made from it and burn them in a weighed platinum dish; weigh the residue, then divide the weight by five.

Swedish filter-paper is considered to contain the smallest amount of mineral matter.

Filter-pump. (See p. 16.)

Flasks, graduated.

Stoppered flasks are preferable.

Those most useful are of the capacities of 1000, 500, 100, 50 c. c. each.

Funnels, filtering. (See p. 15.)

Galena (natural lead sulphide), **Pb S.**

Gauze, iron.

A square of coarse gauze serves as an excellent support for beakers and other vessels that are to be heated over a lamp.

FIG. 42.—Filter-cutter.

Glass covers for beakers.

These covers have the shape of a watch-glass, but are much larger, the convenient sizes being respectively 4, 5, and 6 inches in diameter.

Such covers, used with the convex side down, are serviceable for covering beakers when boiling or when left over-night.

Glass rods.

For use as stirrers.

——— tubing (for apparatus).

It should be of assorted quill size.

——— tubing (Bohemian combustion tube).

A convenient form is in lengths of three feet with an internal diameter of five-eighths of an inch.

Graduates.

One of convenient size holds 60 c. c., and is graduated at each 10 c. c.

Hydro-disodium phosphate (Phosphate of Soda), **H Na$_2$ PO$_4$+ 12 H$_2$ O.**

Hydrometer.

Iodine.

Labels.

For many purposes small gum labels are convenient.

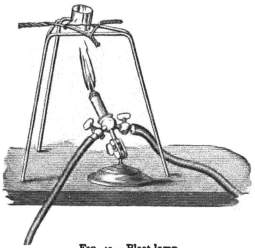

Printed labels for reagent bottles should be of a somewhat porous paper, such as will hold the gum used. (Gum tragacanth is best.) Labels are convenient when made into the form of a book having a very wide inner margin; thus a single label may be cut out and yet leave every other label in its proper position. It is well to have the printed labels separated by a single broad black line; in cutting let

FIG. 43.—Blast-lamp.

the division be at the middle of the black line.

Lamp. (Blast-lamp.)

The blast-lamp is sufficiently explained by the cut.

Lamp. (Bunsen lamp.)

The Bunsen lamp is conveniently made in three parts, the tube, the T, and the base. The tube is of drawn brass. The T is of brass and cast in one piece, as shown in the cut; a clear passage is bored from a toward c, but a thin wall is left at c, this wall being afterward pierced with two minute gas-holes at c; next another gas-passage is bored from b to the previously-bored passage; then a is stopped with a plug of solder. A screw-thread is tapped at a and another at c; the one fits a corresponding thread in the base, the other fits a thread in the upright tube. The base is of cast-iron.

FIG. 44.—Dissected Bunsen Lamp.

In attaching lamps to the work-table it is preferable to have the gas-cocks at the front; if the student has to reach over to the remote side of the bench, he is frequently in danger of overturning his apparatus.

Magnesia solution. (See p. 59.)

Magnesium chloride $Mg\ Cl_2 + n$ aq.

It is crystalline, but upon exposure it absorbs water from the air.

M

Magnesium sulphate (Epsom salt), **Mg SO₄ + 7 H₂ O.**

Marble.   (See Calcic carbonate.)

Mercuric chloride (Corrosive sublimate), **Hg Cl₂.**

Mortar, agate.

—— wedgewood.

Oven, drying.

> Most forms of apparatus for producing distilled water are provided with cells surrounded with hot water.   In these cells substances may be subjected continuously to a temperature of 100° C. (212° F.).

> Sometimes small ovens containing water and heated by gas- or alcohol· lamps are used.

FIG. 45.—Drying Oven.

> A little copper oven (without water) heated by a lamp and having a thermometer serves as a very convenient air-bath.

Paper for filters.   (See p. 15.)

—— glazed.

—— litmus.

Platinic chloride (**Pt Cl₄**).   Its preparation.

> *First method.*   Dissolve 40 grammes of solid Platinic chloride, called Bichloride of platinum, in about 400 cubic centimetres of water.

Second method. Dissolve any weighed amount of scrap platinum in repeated small portions of aqua regia in a glass flask. A convenient proportion is

> 1 part Platinum (by weight),
> 2 " Nitric acid,
> 5 " Hydrochloric acid.

In the neck of the flask place a small funnel to condense and return acid vapors. The metal dissolves very slowly. When the solution of the platinum is accomplished, transfer the liquid to a porcelain dish and evaporate to dryness. Redissolve the residue in hydrochloric acid and again evaporate to dryness, this time on a water-bath. Dissolve the residue in a quantity of water that has 20 times the weight of the platinum used.

Third method. In a clean earthen crucible and over a gentle fire melt 10 parts of metallic Zinc; into the melted metal throw small fragments of scrap platinum little by little to the amount of 1 part. Pour the melted alloy upon an iron plate to cool. Digest the alloy in hydrochloric acid to dissolve the zinc; pour off this solution and then boil the residual spongy platinum in a new portion of the acid. Wash the platinum and then dissolve it in aqua regia. This latter operation must be performed cautiously, as the spongy metal sometimes dissolves rapidly and with violent evolution of gas. When the platinum is dissolved evaporate the solution and treat it as described under the second method.

Fourth method. The precipitate that is obtained from waste platinum solutions may be utilized. This precipitate is generally a mixture of Ammonio-platinic chloride and Potassio-platinic chloride. Weigh a portion of the dried precipitate and then carefully but thoroughly ignite it in a porcelain crucible. (If the operation is performed in a platinum crucible, it is apt to give rise to lumps on the inside of the dish.) The ignition decomposes the salts, leaving metallic Platinum, Potassium chloride, and portions of undecomposed salts. Rub the residue thoroughly in a mortar; then wash it by decantation so as to return the undecomposed salts to the bottle for Platinum waste. Treat the residual metal as described under the preceding methods. One part of spongy platinum requires about five parts of aqua regia for its solution.

Platinum cone.

——— crucible.

——— triangle.

Plumbic nitrate (Nitrate of lead), Pb (NO$_3$)$_2$

Lead sulphide (Galena), **Pb S.**

Potassium bromide, **K Br.**

—— carbonate, **K$_2$ CO$_3$.**

—— chlorate, **K Cl O$_3$.**

—— chloride, **K Cl.**

—— chromate, **K Cr O$_3$.**

—— dichromate, **K$_2$ Cr$_2$ O$_7$.**

—— ferricyanide (Red prussiate), **K$_6$ Fe$_2$ Cy$_{12}$.**

—— ferrocyanide (Yellow prussiate), **K$_4$ Fe Cy$_6$.**

—— iodide, **K I.**

—— nitrate, **K NO$_3$.**

—— permanganate, **K$_2$ Mn$_2$ O$_8$.**

—— sulphocyanate, **K S Cy.**

Potassio-aluminic silicate (felspar), **K$_2$ O, Al$_2$ O$_3$, 6 Si O$_2$.**

Potassio-antimonylic tartrate (tartar-emetic), **K, Sb O, H$_2$ $\overline{\text{T}}$ + $\frac{1}{2}$ H$_2$ O.**

Retort, one of about 106 c. c. (3 ounces) capacity.

      one of about 1 L. (1 quart) capacity.

Sand.

Silver coin.

Sodium carbonate, **Na$_2$ CO$_3$.**

—— chloride (common salt), **Na Cl.**

—— hydroxide (caustic soda), **Na O H** + aq.

        For solutions, dissolve

           1 part white stick soda

         in 10 parts water by weight.

—— tetra-borate (borax), **Na$_2$ B$_4$ O$_7$ + 10 H$_2$ O.**

Stannous chloride (tin crystals), **Sn Cl$_2$ + 2 H$_2$ O.**

Starch, **(C$_6$ H$_{10}$ O$_5$).**

Tartar-emetic. (See Potassio-antimonylic tartrate.)

Test-tube (short), 1 inch long by $\frac{1}{4}$ inch diameter. (See p. 148.)

Thermometer. (See p. 83.)

Thistle-tubes.

Tin crystals. (See Stannous chloride.)

Triangles.

Tripods.

Tubing of glass. (See Glass.)

Tubing of rubber.

Tweezers of iron.

Wash-bottle.

> The ordinary form of wash-bottle seems to answer all purposes. Many modifications have been proposed. (*Chem. News*, Vol. 36, pp. 119, 165; *ib.* Vol. 37, pp. 23, 110; *ib.* Vol. 39, pp. 19, 190, 227.)

> The only appliances we recommend are a rubber joint in the exit-pipe, so as to make it flexible, and a band of cork, wood, or leather around the neck of the flask, so that it may be handled when hot.

Watch-glasses. (See Glass covers.)

> Watch-glasses of ordinary size are used for weighing the substance to be tested.

FIG. 46.—Weights in a Box.

Water, distilled.

Weights. (See p. 37.)

White arsenic. (See Arsenious oxide.)

Wine, spirit of. (See p. 130.)

Wire, iron (for copper precipitation) see

" " (piano-forte), the article used by florists.

———— copper.

Woulff's bottle, capacity 1 quart.

Zinc, metallic.

———— sulphate, $Zn\ SO_4 + 7\ H_2O.$

16

TABLE OF ATOMIC WEIGHTS.

Name of Element.	Atomic Symbol.	Exact Atomic Weight.	Approximate Atomic Weight.	Name of Element.	Atomic Symbol.	Exact Atomic Weight.	Approximate Atomic Weight.
Aluminium....	Al..............	27.0090	27.	Molybdenum..	Mo..............	95.5270	95.5
Antimony......	Sb (Stibium).......	119.9550	120.	Nickel.........	Ni..............	57.9280	57.9
Arsenic	As..............	74.9180	74.9	Niobium	Nb..............	93.8120	93.8
Barium........	Ba	136.7630	136.8	Nitrogen.......	N..............	14.0210	14.
Bismuth	Bi..............	207.5230	207.5	Osmium........	Os..............	198.4940	198.5
Boron.........	B..............	10.9410	10.9	Oxygen.........	O..............	15.9633	16.
Bromine.......	Br..............	79.7680	79.8	Palladium.....	Pd..............	105.7370	105.7
Cadmium......	Cd..............	111.8350	111.8	Phosphorus...	P..............	30.9580	31.
Caesium.......	Cs..............	132.5830	132.6	Platinum......	Pt..............	194.4150	194.4
Calcium.......	Ca..............	39.9900	40.	Potassium.....	K (Kalium)......	39.0190	39.
Carbon........	C	11.9736	12.	Rhodium.......	Rh..............	104.0550	104.1
Cerium........	Ce..............	140.4240	140.4	Rubidium......	Rb..............	85.2510	85.3
Chlorine.......	Cl..............	35.3700	35.4	Ruthenium ...	Ru..............	104.2170	104.2
Chromium	Cr............-....	52.0090	52.	Samarium	Sm..............	150.0210	150.
Cobalt	Co......	58.8870	58.9	Scandium	Sc..............	43.9600	44.
Copper........	Cu (Cuprum).......	63.1730	63.2	Selenium.....	Se..............	78.7970	78.8
Didymium	D..............	144.5730	144.6	Silicon........	Si..............	28.1960	28.2
Erbium........	E..............	165.8910	165.9	Silver........	Ag (Argentum)..	107.6750	107.7
Fluorine.......	F..............	18.9840	19.	Sodium........	Na (Natrium)....	22.9980	23.
Gallium.......	Ga..............	68.8540	68.9	Strontium....	Sr..............	87.3740	87.4
Germanium...	Ge..............	....	72.3	Sulphur.......	S..............	31.9840	32.
Glucinum.....	G or Be (Beryllium)	9.0850	9.1	Tantalum	Ta..............	182.1440	182.1
Gold...........	Au (Aurum)........	196.1550	196.2	Tellurium....	Te..............	127.9600	128.
Hydrogen	H..............	1.0000	1.	Thallium.....	Tl..............	203.7150	203.7
Indium........	In	113.3980	113.4	Thorium...,...	Th..............	233.4140	233.4
Iodine.........	I..............	126.5570	126.6	Tin	Sn (Stannum)....	117.6980	117.7
Iridium........	Ir..............	192.6510	192.7	Titanium.....	Ti	47.9997	48.
Iron...........	Fe (Ferrum)........	55.9130	55.9	Tungsten.....	W (Wolframium).	183.6100	183.6
Lanthanum ...	La..............	138.5260	138.5	Uranium......	U..............	238.4820	238.5
Lead...........	Pb (Plumbum)......	206.4710	206.5	Vanadium....,	Va..............	51.2560	51.3
Lithium.......	Li..............	7.0073	7.	Ytterbium,,..	Yb..............	172.7610	172.8
Magnesium...	Mg	23.9590	24.	Yttrium.,.....	Y	89.8160 '	89.8
Manganese.....	Mn	53.9060	53.9	Zinc.,........	Zn........	64.9045	64.9
Mercury.......	Hg (Hydrargyrum).	199.7120	199.7	Zirconium....	Zr	89.3670	89.4

INDEX.

183